To Mary F
With Vitamins, Gordon Baxter
April 1996

More
BAX Seat
New Logs
of a Pasture Pilot

Other Books by Gordon Baxter

**The Al Mooney Story
BAX Seat: Log of a Pasture Pilot
Best of BAX: Collected Titles from** *Car and Driver*
**How to Fly
Jenny 'n Dad
Village Creek
13/13 Vietnam**

More BAX Seat

New Logs of a Pasture Pilot

Gordon Baxter

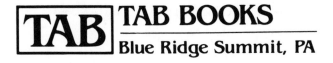
TAB BOOKS
Blue Ridge Summit, PA

Portions of *More BAX Seat: New Logs of a Pasture Pilot* originally appeared, in somewhat different form, in *FLYING* and *EAA Daily*.

FIRST EDITION
SECOND PRINTING

© 1988 by **Gordon Baxter**.
Published by TAB Books.
TAB Books is a division of McGraw-Hill, Inc.

Printed in the United States of America. All rights reserved. The publisher takes no responsibility for the use of any of the materials or methods described in this book, nor for the products thereof.

Library of Congress Cataloging-in-Publication Data

Baxter, Gordon, 1923-
 More Bax seat.

 1. Private flying—Anecdotes, facetiae, satire, etc.
2. Flying (Chicago, Ill.) I. Title.
TL721.4.B35 1988   629.13   88-8574
ISBN 0-8306-9029-8
ISBN 0-8306-9429-3 (pbk.)

TAB Books offers software for sale. For information and a catalog, please contact TAB Software Department, Blue Ridge Summit, PA 17294-0850.

Questions regarding the content of this book should be addressed to:

Reader Inquiry Branch
TAB Books
Blue Ridge Summit, PA 17294-0850

Contents

Introduction ix
1 • The James River, So Dark and Wide 1
2 • Super Stinson 8
3 • Me and Elmer Lee 12
4 • Beyond the Hill 16
5 • Flying the Space Shuttle, Simulated 19
6 • Thigpen and the Landing Gear Handle 25
7 • NBAA Convention 30
8 • Blup in the Night 34
9 • The Texas Gunner 38
10 • Approach Wants You on the Phone 42
11 • When the Tri-Pacer Was New 46
12 • Skydiving Cats (Part I) 50
13 • Oshkosh Tower 53

14 • Eleanor   57
15 • Merced   62
16 • The New Boy   67
17 • Justifiable Scud Running   77
18 • Skydiving Cats (Part II)   81
19 • Braniff's Final Run   85
20 • Sea Stories   91
21 • About Spins   97
22 • Barnstorming at Port Arthur   103
23 • Jeana's Eyes   109
24 • The Ten Best   112
25 • What Did You Do in the War, Dad?   119
26 • All Chuck, No Yeager   125
27 • Skydiving Pigs   130
28 • Concorde   134
29 • Hello Stearman, Old Friend   136
30 • The Past Fifty Years   140
31 • The Great Mooney -
     vs. - Boeing 727 Race   146
32 • What Shall We Do with Ellington?   151
33 • That Big Purple Airplane-Eating Cloud   157
34 • Super Connie   163
35 • Like Torpedoes Being Loaded   169
36 • Testing the Tires   174

37 • A Day in the Life of a Girl Watcher   180
38 • Grounded   185
39 • Epilogue   185
   SEMI-GLOSSary   190

Introduction

More than just an aviation magazine, over the years *FLYING* has almost become an institution, cruising its stately course with an overview of aviation. The magazine is seldom mean with criticism, seldom swept away with praise. There is, however, a surprising depth of research and sense of responsibility which results in *FLYING* being respected and trusted.

For such a large organization the magazine is run with a remarkably small amount of inside strife. I am one of the many far-flung writers. My relationship with the magazine is very adult. If a story is good, they send the check. If not, they send it back.

My one-page column traditionally appears on the back page, thus the punned title, "BAX Seat." In the past decade and a half I have come to think of myself as *FLYING*'s story teller. It is almost as if the magazine says to you, "There, we've taught you all we can this month. Now here's Bax. Enjoy."

The first book of *BAX Seat* was originally published in 1978, so the material here is from that date forward. These are the stories I considered as keepers, the ones worth telling again.

1

The James River, So Dark and Wide

ALL IT SAYS in my 1962 logbook about this is "Spiraled out—IFR weather." I was afraid to put anything in there about the spin, but the horrifying details will live in my mind forever.

I was a private pilot, heavy with a total of 250 hours of flying time, but had never really been anywhere yet. I did what any Texas boy would do: decided to fly to New York City.

We were not burdened down with a lot of stuff in those days. Nosewheels were a recent development and looked like they might open flying to the inept and slow-witted. Omni navigation devices were starting to appear in more and more airplanes, but they came after I had completed my formal instruction. Nobody mentioned them, nor was I overly curious about them.

I couldn't see how a lot of the stuff they taught applied to me. Like all that froo-frah about figuring wind drift on the backside of the big plastic computer. You could spend all morning doing that; then, by the time you got into the airplane, the wind had changed. I just drew a line on the map for where I wanted to go, found a good straight stretch of road or railroad that lay along my line and followed it, noting how I had to crab to stay on the line, and sailed along. I got there just the same.

Another thing they often warned of, and I read a lot about in the aviation magazines, was "Don't fly in the clouds."

I didn't, but it sure was hard to believe all that stuff about spatial disorientation and not being able to know which way was up, not just sitting there flying along. I always took a little bust at whatever puffball clouds were in my way. If being in clouds was all that bad I wanted to know at least a little about it. Bump, woof, and out again, level as a table. Seemed to me if a pilot were in one of those things he could take off one shoe, hang it from the throttle by the laces, and just fly "shoe down."

I was flying a beautiful old Cessna 150 and made it to New York City in a couple of days, just cruising along for the sheer fun of it. I spent one night sleeping in a hangar at Macon, Georgia. I shared the hangar floor with a big old wooly dog who curled up with me that night, and we kept each other warm.

After a total of 15 hours of truly beautiful flight, the great grey shadowed wall of the ramparts of New York City filled my windshield, and I decided that was close enough, saw little airplanes parked at Linden, New Jersey, and landed there.

Headed homewards the next day I saw the big airport at Baltimore, gave it lots of room, and heard them tell an approaching DC-7 of a "slow moving target." I guess that was me. Slow moving yes, target never. I thought that was kind of funny and flew along laughing about it.

I landed at a little airport at Washington, D.C., they had several back then, and phoned the FSS about the lack of sunshine ahead.

The weatherman advised me of a frontal system moving in from the Great Lakes that would soon close the sky for a few days. I told him my plan to fly the Shenandoah Valley out of there. "It might be a little bad through there, but Knoxville will be clear."

It was a little chancy, but I had read a lot about making a 180 and getting the hell out if things got too bad. I took off; the beautiful Shenandoah Valley went right in the direction I needed to go.

What got me was something I had never even heard about: the meeting of the dew point and temperature. My little homeward valley was starting to fill with lovely deadly white fog in the low places.

I had been flying beneath an overcast for some time now, lower than the mountains on each side. I made a careful 180 and stared into a solid wall of white. The lowering ceiling and rising fog had silently shut and locked my escape hatch behind me. I had never even thought about it doing something like that.

I made another careful 180, just sort of hoping I would reach the southernmost edge of this before it got any worse. "Bax, you are boxed." This was no time to laugh, but talking to myself helped some. I had to fly lower and lower and slowed down so the barns and tall trees were not appearing too fast for me to dodge them. I was dodging farmers on tractors, had the image of them shaking their fists at me, and got to laughing again. I wasn't really scared yet. But I had run out of options. That is, all but finding a place to set her down very soon.

The Shenandoah Valley is truly beautiful country but it sure ain't like Texas when you have to find a place to park an airplane. All hills, even the hills had little hills. And trees and rocky outcroppings everywhere. This was where I found out how hard it is for a pilot to consider deliberately wrecking a good-running airplane, even if it means saving his neck. I knew I could stall-in that old straight-back 150 (with full flaps at about 40 mph) and hang it in a tree somewhere, but I just couldn't do it. The valley grew darker and darker.

Then a circle of light appeared ahead. I could turn tightly inside of it, look up, and see blue sky. That was when I found out you can't make an airplane climb when you are turning it as tight as you can. I went round and round in my little circle of life, trying to rise, like a fish in a tank, gasping out his life. I could feel it when the airplane was going to quit on me in the turn and never pushed her too far.

I rolled out of it, back into the dank darkness, back to barn and hill dodging again, only now my compass was swinging wildly, and I wasn't really sure of whether or not I was flying out of it, backtracking, or headed across the valley and into the shoulder of some mountainside. That was aggravating.

The little whiffs overhead had shown me how thin the overcast was. From that I made the next decision. To fly up through it. Surely I could last two minutes in clouds.

I took a deep breath. An inner voice that seemed distant—not passionately involved in this—said, "You are making a life-and-death decision, you know that don't you?" I said I knew, sat rock-steady in the seat, pushed the throttle wide open, came back gently on the yoke, established a 90-mph climb, and whoosh, we vanished into the cloud layer.

It was about what I expected. I just sat there listening to the engine, holding still. In those days they taught a student nothing about how to read the horizontal situation indicator or any of the rest of it. About a minute passed and it felt like we were still straight and climbing. "I'm getting away with it" was my thought.

Then it began to get more light inside the cabin. Not daylight yet, but a thin glossy white. I knew I must be at about the top of it. I leaned my head back to peer up through the top of the windshield. That did it.

I had no way of knowing what happened, but from the increased sound and the unwinding of the altimeter I was pretty sure I was in a spin and didn't know which way it was turning. I knew I had just a minute or so left to live, but felt no real fear then, no panic. Just sitting in the balcony watching my own movie run out.

My mind kept searching for something that would be useful, some way to do something about this. Then I remembered my instructor telling me about spins (they taught spins back then). "They build very stable airplanes in Wichita. This plane does not really want to spin. You have to force it. Turn everything loose and see what it does." The airplane came out of the spin by itself and entered a dive.

"I sure hope that works this time," I was thinking. I pulled the throttle back to idle. I was still calm, almost dispassionate about it. I even remember reaching down and pulling out the carb heat knob, too. I crossed both arms over my chest, moved my feet back from the rudders, and began to watch through the windshield for the first dark smear of green that would tell me I was at the tree tops. I could tell by the gentle sound of things that I was no longer spinning. The next question was: would I come out of the clouds with enough room to pull out of the dive?

I broke out just over the wet tree tops in a gentle gliding left turn. "Son of a gun! You made it!" I eased the yoke back to level flight. Then the engine quit.

In all my preparations before dying I had pulled the mixture control instead of carb heat. I slapped it back to full rich, and glided, steering between the tops of two tall trees before the engine roared back to life.

Back to tree and barn dodging, only now I was just looking for a soft easy place to crash it. I had no hope of flying any farther.

Then, just ahead, I saw a long trough of light, a long wide slot with sky overhead. I got into it, flew to the end, made a steep pylon turn, stayed inside the light slot, climbed toward the other end, made another pylon turn at the wall, and flew the slot, climbing at max power. I kept doing this, like going up an escalator in a six-story department store, then popped up between cloud layers. Enough sky, enough horizon to live.

"Bax, the Lord just don't want you yet," said my other voice inside. I waited until the compass settled down, then set up a course that I hoped would be toward Knoxville and the edge of this stuff.

The layers got wider apart. One of the layers was inclined, and I noticed that even with a good, flat horizon of clouds ahead, I tended to lean toward the slanted layer. I was now a convert to the truth about everything they ever warned of in-cloud flying, and there was a whole lot about it I still didn't understand. A much more sober and penitent young pilot, I flew on into clearing skies, found Knoxville, and the weather seemed good enough to go on and fly to Atlanta. I had lots of time to sit and think. Feeling strangely calm, I didn't sing in the airplane. Lots of times alone I sing in the airplane. Now I just flew towards Atlanta, staring.

I headed for the big airport at Atlanta because it was easier to find. Hartsfield was a big airport back then, but not nearly as soaked with traffic. It looked green and welcoming below, a few cloud shadows drifting across the broad runways. I found the tower frequency on my sectional, turned the little crank on my radio until I heard a tower talking, and announced my intention to land. They gave me a runway number. I didn't write it down and instantly dropped it out of my mind. The landing was coming up too soon to be fooling with the radio again. Tower called me, asked if that was the runway I intended to land on, and gave me two numbers for it. I just told them yes, but had no idea of what runway it was. They had lots of runways; none of them were busy. I wasn't about to tell them I had chosen this one because it was the first one I found in the direction I was going.

It was a weird landing. I put it down to the visual effects of never before having landed on a runway so wide and so long. I took pride in landings, even then, but really had no idea of how close I was to this one. I just landed on it the same way I would land in the dark with no landing light. The Cessna accepted that with grace.

I found I could turn off at the first intersection and headed toward a hangar where I had seen little airplanes parked. The tower was talking to me again, but again I couldn't remember any numbers they were saying, and just left the radio alone.

I parked in a row of other Cessnas and Pipers, got out, and left the airplane. I had enough, and I knew it. I was looking for a place within walking distance to spend the night. I couldn't find a place to buy whiskey but I did find a little motel and went into my room and sunk into the hottest bath I could stand.

Floating in the hot tub, chin-deep, the horrors began. All the old train-wreck ballads began to run through my head. Where I had been, near Roanoke and Lynchburg, was where Steve wrecked old 97. And there was another where the engine, speeding along the curving track that edges the James River, hit a rock and rolled into the river, ". . . so dark and wide." They found poor Harry Lyle in the cab ". . . with a deep fatal wound in his head."

I remember flashing across a river I thought might be the James while I was lost. For an instant I had considered doubling back and trying to find it again to see which way the current was running, but just dodging things straight ahead was all I could handle then.

The front page of the Beaumont Enterprise flashed before my eyes there in the tub, "Baxter Crashes Into The James River So Dark And Wide . . . ."

I opened my eyes, and the tub was filled red with blood. I closed them again, and the wall of the bath opened into a view of that forest by the river. There were bright shards of aluminum hanging in newly torn trees and scattered along the ground. All was silent except for dripping sounds in the fog. The little Continental engine was stuck into the ground, and my intestines were strung out over the still-hot engine.

I arose in a rush, fled the bathroom, grabbed for my clothes, and walked unsteadily toward the little motel restaurant. There was one empty seat at the counter, and when I sat down, the persons on either side seemed to be unaware of my being there. They neither looked nor moved. I watched the waitress walking back and forth behind the counter. She too was unable to see me.

Which was real? Was I actually dead back there beside the James River, and this was just my last dream of how I hoped it would work out, or was this real and the gory crash beside the river was a dream?

That plump, pretty, little Georgia girl will never know how much it meant to me when she stopped at last and asked what I wanted. I nearly asked her to marry me, but ordered coffee instead.

Although long since instrument-rated, I have never really been happy flying in clouds, and the details of that day have diminished little. There were some things I was still curious about, and years later I asked a neurosurgeon friend about the seeming lack of fear at the time, the inability to remember numbers from the Atlanta tower, and the dislocated feeling of that landing.

"Some people have a controlled level of panic. The higher brain center has already quit, but the simpler, learned, motor-control functions are still available. You could fly and land pretty good, but couldn't handle being given two new numbers. The state you were in comes just before the screaming stage and lasts until after the emergency has passed. Sustained for extended periods of time, it was once called 'shell shock.' The experience you had didn't last long enough to make you any more squirrelly than you always have been."

I told him that sometimes it's great to have friends in the medical profession.

2

Super Stinson

NEELY JOHNSON WAS MY AIRPORT HERO back in the mid-1950s. He kept the most interesting airplanes: a tiny little single-seater called the Mooney Mite; a "turkey tail" Bellanca, which is the undeserved nickname for the tail-dragging, three-rudder, first model of the low-winged Bellanca; and a Stinson Voyager with a big engine in it, which he called the Super Stinson.

Neely had a fine mind for the advanced ideas of aviation back then. We would stand together in the echoes of the big hangar at Jefferson County Airport between Port Arthur and Beaumont, Texas, and he would tell me of the advanced features such as the laminar-flow wing and retractable gear of the tiny yellow-and-black Mooney, and of the lift curve which was said to be the reason for the humpbacked appearance of the old Bellanca.

Neither of these engineering masterpieces were in flying condition; they sat coated in dust on low, cracked tires. Neely had great plans for the restoration of both, yessir, gonna make them like new. Someday. Soon.

Meanwhile, things kept falling through the Bellanca, and the tiny Mooney looked like somebody in big shoes had been stepping on it. Gaping black holes kept showing up in both airplanes. Neely would look at the damage, stand back, and cluck and shake his head. For a man who strived towards perfection and appreciated these two hulks for what they could be someday, the hangar damage struck at his very heart.

But he did have one airplane that would fly, one that satisfied his love of things made well and, perhaps, differently. This was his Stinson Voyager, which he said was a Bonanza from the firewall forward.

All-white, with a dark-green trim stripe, his "Super Stinson" stood on speedy-looking wheel pants. I don't know what happened to its original Franklin engine, but its long alligator snout, raised skyward, hooded a Lycoming of 190 horsepower. In those days anything over 65 hp was a big engine.

The only outside clue that this was no ordinary Stinson was the propeller with a chunky-looking metal hub and wooden blades. Neely said this was a "two-position" propeller whose pitch, controlled from the cockpit, could be shifted from a clawing climb pitch to a high-speed cruise setting. Oh, I had to fly that airplane.

My old logbook shows that Neely gave me an hour-and-a-half dual in it on November 30th, 1956. That brought my total time up to about 56 hours and satisfied us both that I was ready to move up to this high performance airplane. He warned me to wheel land it only, that the weight of the bigger engine up forward had done something funny to the original balance of the Stinson. Yes, it had.

With most of my time in Champs and Luscombes, the Stinson felt like a big airplane—the big yoke in my hands, all that power—and wheel landings were easy in it. Like loading a canoe, there was lots of leverage, so you knew when it was level.

On the very next day after checking out in the Stinson, I launched for Mobile, Alabama, 362 miles away, my longest cross-country at that time.

I did dead reckoning navigation, sort of, but was never sure of it. Trouble was I would get to sightseeing out the windows, daydreaming, and enjoying flying too much, and not pay attention. And there was no radio in this airplane. It apparently had caught fire recently and had been shoveled out. There was only a charred hole in the panel where the radio used to be. Neely and I didn't discuss this.

Small loss. Fly Highway 190 straight to Baton Rouge, then follow the Mississippi down to New Orleans, then pick up the Gulf coastline over to Biloxi, then turn left to a heading of about sixty degrees for the short trip inland to Mobile. I ought to be able to do that.

I was excited at the start of this trip. The big engine sounded a mighty roar; my ears were laid back at the thought of passing over the ground at a rate of two miles per minute. I quickly felt and appreciated the rock-steady, ball-bearing feel of the Stinson aircraft. I even wore a clean white shirt and coat and tie to be worthy of such a ship.

Somewhere in all this new glory I noticed the airspeed indicator had lapsed quietly into its corner, never to stir again. No problem. I was used to flying airplanes by the feel and sound. I made a DC-3 approach into Ryan Field at Baton Rouge, a good wheelie, and never really missed the airspeed indicator, not then nor later.

My log shows I bought 21.8 gallons of gas for 31 cents per gallon and two quarts of oil at 40 cents each. Standing in the shade of the big Stinson, I was treated with respect.

The flight down the mighty Mississippi was sheer beauty, as was flying over the crescent city of New Orleans, then along the Biloxi gold coast with all its fine resort homes, white sands, and clear waters below.

Soon I was at Biloxi, time for the cross-country stretch up to Mobile. The heading was 60 degrees, and I was amazed at how rock-steady I held it. Landing at Bates Field, I noticed the compass was still rock-steady at 60 degrees. I watched it as I made the turnoff toward the ramp. Still sixty. It was stuck or something. Sixty degrees was all it had on it. Never heard of an alcohol compass doing that before. I decided if I could get there, I could get back, and put the compass out of my mind.

At Mobile the Stinson took another three quarts of oil. A little concerned about using five quarts in three hours flying time, I got down and looked under. The entire belly was thickly coated with oil, a clear golden color. Whatever that Lycoming was doing with all the oil, the oil wasn't getting very dirty.

Old friends Joel and Marilyn Swanson met me for the overnight at Mobile, and still elated from getting there, I offered them a ride over the city. Joel looked embarrassed trying to find gentle words to tell me no. Little Marilyn eased the moment for both of us by opening the cabin door and sniffing inside. "This thing smells like you cooked your breakfast in it." There was laughter all around, and we were off for a pleasant night with the Swansons.

Getting a green light from Bates Tower, I joyously looked forward to a day of flying back home. Nothing to worry me, not the compass, the radio, not the airspeed indicator. Good highways to follow all the way. But a deep shuddering seemed to wrack the Stinson as I changed the propeller from climb to cruise pitch. I was later to find out that only one blade of this propeller was changing pitch. There was, as Elvis used to say, a whole lot of shakin' goin' on. And I began to smell gasoline.

Now, all the old airplanes always smelled a little like gasoline, but a lot of wind blew through the cracks in the doors. The airplane took four more quarts of oil at Baton Rouge, and I began to worry some about the engine. The gas smell was getting stronger, too.

At one point, I decided to ease my fears and bring back the fine mood this trip had begun with, and leaned back and took out a slender little cigar. I held that thing a moment, savored its aroma, but something seemed to tell me, "Bax, don't strike a match in this airplane."

When I wheeled it in, back home at Jeff County, I noticed the underfabric of the wings sagged heavily and little beads of gasoline were dripping off of everything. It seems the vibration had cracked both gas tanks, and I was flying what may have been the world's only wet-wing Stinson.

I carefully put the unlit cigar back in my pocket, tiptoed away from the airplane, and went in to tell them about it at the airport office.

Would you believe they not only grounded that airplane but they cordoned it off so nobody could get near it? The airport operator said he thought the Lord was obviously saving me to be hanged. I told him nobody born on Christmas Day was ever hanged.

Neely finally sold that Super Stinson to his good friend, Dr. George Sims, who restored it to immaculate condition. I bet that beautiful deep-bodied, high-tailed Stinson is still flying somewhere today, filling the heart of some pilot with its might roar.

3

Me and Elmer Lee

WE BOUGHT OUR FIRST AND ONLY AIRPLANE on March 17th, 1975. It's a little orange-and-white 1968 Mooney Ranger, Model 20-C. One of the happiest of airplanes, one of the best of the old Mooneys. I flew it free and far for a year and a half before a little glitch in my brain took away my medical, probably forever. For those of you in the medical trade, it was a calcified AVM in the left occipital. No threat to life, I'm still healthy as a horse, but I do sometimes have very brief blackouts. The FAA is very picky and absolutely one-way about their idea that the pilot must be conscious at all times. I didn't even apply for another medical when my current one ran out. No use opening a case with the feds when I don't have a case. If by some miracle this condition goes away, I will just apply for a new medical as if I had only been gone a long time. There will be no case in their computer for them to chew over.

    No secret that I had lost my medical, a young couple, Elmer Lee Ashcraft and his pilot-wife Elaine came to call and asked if I wanted to sell the Mooney, No way. Just no way could I part with that little dart-tailed time machine. Better I should sell them one of the kids. I hemmed and hawed and asked if they'd like to buy half of it. An airplane needs to be flown. I was still flying then, as now, whenever I could find a licensed pilot to go with me. I took a liking to the Ashcrafts from the first time I ever saw them. Both slight built, both sort of redheads, they looked like brother and sister and looked like they could have been part of my family, too. And anyway, he is a CFII. The airplane would get flown more, and I'd have a partner I could lean on for me to fly.

Like marriages, there are good airplane partnerships and bad, and a fellow has no way of knowing which it will be when he's taking the vows.

Our partnership, which began on May 3rd, 1978, still exists but has grown into a warm family friendship, more than just airplane partners. The wives like each other, the Mooney likes Elmer, and so do I. In the past eight years since we became partners with Elmer and Elaine, we have had our economic woes. Sometimes sat across the coffee cups at a midnight table and seriously discussed selling the airplane. By the next morning all had agreed we just couldn't do it. But our faithful Lycoming 180 had run out its time, and we needed thousands for a new engine.

That was when we decided to take in a third partner, hereafter known as the "engine partner." That brought David Shelby into our little family, and as before, none of us ever stood on the other's toes or got in each other's way.

When the Arabs began to play with the price of oil, the oil town of Beaumont, Texas, spiraled down into deep recession. The Mooney's life with us was threatened again as our own incomes began to shrink. Again we decided we could stand another partner, and the fourth partnership in the old airplane has passed from hand to hand a time or two as fates and fortunes would move one person out, another in. We still have the airplane, but in my heart of hearts I can't believe that I'm no longer a licensed pilot and that I only own a fourth of old "27 November," which is as close as we've ever come to finding a name of affection for Mooney N6727N.

I still think of it as "my" airplane, still think of me as a pilot. It's my head. I can screw it around if I want to.

The first homecoming of the truth was one summer evening about sunset. Driving home on Eastex Freeway out on the edge of town, I was looking at the reddening sky and thinking how nice it would be to be up there flying in that, the sort of thing pilots in cars are always doing. Suddenly, out of the corner of my eye, I caught a distant moving silhouette of an airplane, black against the flaming western sky. I looked again; it was the unmistakable shape of a Mooney. I pulled off the road, got out, and watched. She was so beautiful, the distant drone of that healthy Lycoming, moving toward the airport. I watched, and tears began to come down my cheeks. "Lord, I didn't need that." I gritted to myself, then criticized myself for wallowing in this "why me, poor me" stuff I had not done up to now. Why couldn't it have been a Cessna or a Piper? Why did I have to see my Mooney flying against the sunset? And how did I know it was "my" Mooney? It's not the only one in town anymore. "Bax, you are being an old fool."

I made a U-turn at the next turnaround and, in the gathering dark, began to drive fast out to the airport, also on the edge of town. Wanted to see her come in. Wanted to watch her land. "Knife in your heart is not enough? You want to twist it?" asked the other voice in me, the one that stays sane most of the time.

I parked with a good view of the ramp and so I could also see the approach end of the active runway. The Mooney was banking gently into final, landing light bright, strobes flashing. Now I knew it was 27 November. She settled from the red glare of the nearly gone sun into the dark of the earth shadow, flared delicately, and was swallowed up in the night. I could see by the position of the lights how well she had been brought in. "That's got to be Elmer Lee. He lands like that."

The little plane was taxiing toward the ramp now and would pass right in front of where I was parked. Elmer had all the consumers on: strobes winking, collision light turning, nav lights bright, landing light moving gently up and down over the irregularities in the ground. Making a regular circus of it. Both of us knew how festive she looked with all her lights on, taxiing like that on a deserted ramp.

As she passed directly in front of me I saw they had the cabin lights on, too, and recognized them sitting in there so cozy. Elaine was moving; she would be stacking all the charts away now, packing up the stuff. They are neat people. Fly neat, leave a tidy cabin.

I watched our little airplane go bobbing on by, little white taillight trailing off toward our T-hangar. Oh, they looked so fine in there.

Did I say "our" airplane? Is it possible I am getting over it some? My first impulse was to follow them in the old pickup and help Elmer do the push-back into the hangar, but no, this was their moment, and anyway, I wasn't sure of how my face looked yet.

They are both keen and sensitive people. What if one of them looked at me and said, "Gordon, what's the matter?" Or what if they didn't?

Our Mooney bobbed on down and around the corner of the row of hangars. I got back into the pickup and took a long deep breath—some might say it was more of a sigh—and drove off towards home.

A few days later a flying preacher friend called and asked why I was staying away from the airport. I told him, and he seemed surprised.

"Different people handle it in different ways," he observed preacherly. He said he was surprised I wasn't just flying anyway. "You've always lived like the rules were written for somebody else."

I told him the rules were written to protect the slow. From each other.

He laughed and told me the story of a man about my age who had gone in about a year ago. This man was flying without a valid medical, too. When the investigation was over the insurance investigators had come, patted the widow on the head, and said to her, "Not a dime."

The preacher also said, "Don't get into playing 'Poor Me'." I told him I was trying not to and doing better all the time. That dual flying is better than no flying at all. But solo flying is like being in the bathtub. It's best by yourself, and if you gotta have somebody in there with you, you get picky about who.

The preacher laughed and said he thought I was going to make it. I told him I thought so, too, but I wish there had been some way I could have known the meter was running out on solo flying. I sure wasted a lot of days when I could have gone flying and didn't.

4

Beyond the Hill

IT WAS EASY AND NATURAL to get back into using the airplane again and to include Elmer Lee and Elaine in the trips. I was going to Alliance, Ohio, to do a story on how they make Taylorcrafts (one at a time). The first stop on the trip was to buy gas at Nashville. It was the summer of 1978.

Nashville had not yet gone IFR, but they just about should have. It was my leg to fly, and I was talking to Nashville Approach as they radar vectored us through the thickening haze and down into that pocket in the mountains where they keep the airport.

They had kept us up at 4,000 feet until we were about three miles out, then the controller showed his savvy of Mooneys and asked if I would need to broaden the approach to land. I told him thanks but no, and with Elmer Lee sitting right there beside me grinning, I began one of my famous dirty Mooney approaches: flaps down, gear out, and sinking like a stone.

The only thing was my being a little rusty for recent flying and shying off from the runway about ten feet high. The arrival really wasn't all that bad; it just sounded like the airplane was being dumped off a truck. Elmer Lee got the circulation going again in his knuckles and started to snicker. I told him somebody had come out and lowered this airport about ten feet while I had been away from flying.

He flew the next leg, and I told him we'd soon see how good

he was, too. Alliance, Ohio, showed 2,000 feet of turf. And the first problem was to find Alliance, Ohio. We got into the neighborhood of the place, using cross vectors from our own VORs, and I'll bet that every field in that part of Ohio is either a little turf airport or looks like one.

By matching up power lines and highways, we found one that Elmer Lee confidently announced was Alliance. We descended into a low final approach to fit through a notch they had cut in the trees, and I thought then that airport sure looked short and stubby for 2,000 feet.

There was a thrill in the tree notch. When you get close enough you can find the power lines running through it. The burst of power messed up his right-at-the-edge-of-the-grass planned touchdown spot, and he gunned it for a go-around.

Next time we found out the strip went downhill at the approach end. Elmer had a little too much speed on, and the grass kept settling away faster than we were coming down to land on it. Running out of airport again, Elmer elected for another go-around.

The third try was better but carried us past what I judged to be midpoint in that short little strip. I remembered the flash of a solitary man back by the buildings, standing open-mouthed, just watching.

Elmer was committed to land this time, and the downslope had become a hillside. The nose was considerably up and all of us were focused on the hill. It rose rather steeply before us, ending in a sky that was empty except for a marching line of those big man-shaped power-line towers. Big ones that you could hang an airplane in for keeps.

We contacted the hillside turf smoothly, doing about 60, and then rushed up the slope of that hill toward the end, leaving long furrows from brakes in the grass.

It got real quiet in the airplane, each of us occupied with our own thoughts of what lay Beyond The Hill. I felt no real fear. I knew it would only be one of those long, messy crashes that bounces through the ditch and wipes off the gear on the railroad tracks then tears the wings off in the trees. We would all slowly climb out of it, and I would slowly walk back down the hill to the man who would still be standing there waiting, and I'd sell what was left of the Mooney for spot cash and buy all of us tickets on the bus back to Texas.

Elmer's wife, Elaine, sitting in the back, said she was having a vision of our friendly insurance agent laughing and tearing up our policy in front of us. She alone had remembered the "no turf" clause in the policy as we went speeding along through our last few seconds to whatever was Beyond The Hill.

Elmer Lee claims he was still thinking of a way out of this, such as a Smilin' Jack maneuver by locking the brakes on one side, ground-looping until we were going tail first, then using the throttle to stop us at the top of the hill, which to him would be the edge of a big, deep, rock quarry.

The Mooney swept triumphantly to the top of the hill, then mysteriously came to a soul-saving stop. I will always believe there was some yet-unnamed law of physics in play there, that a body in motion can be stopped by three other bodies in motion, rearing back the other way against the motion. And also the cosmic-ray resistance of three sets of eyeballs, magnified as big and round as coffee cups.

We stared to see what was Beyond The Hill.

It was the other half of the landing strip, that's what it was. There was still enough smooth turf ahead of us to land a DC-3. The airstrip just happened to have a little hill in the middle that couldn't be detected from the air by newcomers.

The tiny Mooney cabin, smelling of galvanized iron, was filled with our whoops and guffaws of relief. There was foot stompin' and back slappin' as we taxied back to the solitary figure of the man who was still standing there, now grinning.

Friendly, he told us this was not Alliance, but the famous Koons airstrip, home of the hill in the middle and power lines in the trees. Alliance was just over yonder ways, he pointed.

Now, my suggestion to Mr. Koons is that he organize a big fly-in, offer huge cash prizes to the attendees, but keep the hill a secret. Then, as each pilot comes taxiing back down the hill, bathed in this own juices, offer to park his plane for him, and sell him a front-row seat to watch the next guy come in. I would pay a hunnert dollars to see a Bonanza pilot's face after he's played Beyond The Hill.

5

Flying the Space Shuttle, Simulated

THE PUBLIC INFORMATION OFFICE at the Johnson Space Center, Houston, said all the astronauts were learning to fly the shuttle and the simulator was booked solid. But if I could get there early, before 8:00 A.M., I might get some pilot-seat time in it while it was being warmed-up and programmed for the day. This was in the summer of 1979, before the shuttle was to fly.

I have seen some big simulators, and I have seen some warm-ups, but I have never seen anything as big as this warming itself up so wildly. It looked like a sawed-off mastodon doing fast knee-bends and push-ups. It's a life-size replica of the shuttle, crew section to nose, standing on shiny, scissored legs which act as hydraulic pistons. It filled one end of the two-story simulator room (a computer control center with rows of operators in front of their picture-tube panels), and rows of lights and control switches filled the area behind it at floor level. Looking like King Kong doing a disco, with hoses and cables swinging freely off its back and belly, it was in operation when I crept into the room.

Nobody was paying any attention to my being there, so in a loud clear voice I announced I was scheduled to fly before 8:00 A.M. The technicians turned from their consoles, looked, and shrugged. The shuttle simulator stopped heaving and hissing, and a slat-thin Navy commander in a flight suit opened up a door on the flight-deck level.

He was Pilot in Command.

What happened next has been the basic makings of many an air disaster including this one. He assumed that I had been checked out and briefed or I wouldn't be there. I assumed he was going to check me out and brief me. We were both wrong. So the total amount of my indoctrination before laying hands on the most sophisticated $60-million simulator ever built was somebody saying "Git him a helmet." I wasn't about to admit I had just wandered in off the parking lot and was lucky to have found the simulator room by myself.

I climbed the stairs to the door the commander had left open and peered into the dim light. The cockpit looked a lot like a Lockheed 1011 except it had twin control sticks instead of yokes, and the windows were narrow and fort-like, while the Lockheed jumbo transport had lots of clear forward view in its nose. But then the Lockheed does not have to re-enter Earth's atmosphere wrapped in flames.

I settled down into the right seat, feeling very privileged, excited, and confident. The commander was already in his left seat. True, I had no idea of what I was doing, but I thought, "Give me a few minutes to get the feel of it, and I ought to be able to fly this thing." The historic attitude of the American aviator.

I thought we were going to simulate being dumped off the back of a 747 and then shoot touch-and-goes at Edwards in the desert, but no, we were going to simulate being launched at the Cape, entering Earth's orbit, and then land at Edwards. The countdown had started. From inside it was absolutely real.

There was a large, green, glowing, visual display screen on the instrument panel, and as the launch began, it showed a straight-up pathway with a pitch-over at the top for orbital insertion. The Navy guy pointed to a darker green bug climbing. "Keep that between the lines. You got it."

Well, a lot of funny things began to go wrong here. I had planned on following him through on control movements but he hadn't made any. The stick was about as big around as my leg, and bolted solidly to the floor. Only the hand grip on top moved, like the stick on a Bede 5. All it took to fly this thing were wrist movements.

Then a shudder passed through the spacecraft, followed by deeper shudders until the shaking was coarse enough to slam us side-to-side in the shoulder harnesses.

"Entering sloshing," said Control crisply in my earphones. I nodded as though I understood.

"Sloshing," I later found out, is what the computer predicts will happen during the launch when about half the fuel is burned off. Keep in mind that the main fuel tank is many times larger than the shuttle. We were tacked onto it like a mockingbird on a tank car going 3,000 mph. The sloshing stabilizes itself just before orbital entry. Normally, the crew would just ride through it, but nobody would really know until the thing was actually flown. What we are talking about was only a computer forecast before the first flight.

Fuel-consumption numbers here are hard to imagine. With solid-fuel rockets pushing from each side, the three main engines in the shuttle body use about a million-and-a-half gallons in the first eight minutes. Then you are out of gas, the boosters and fuel tank drop away, the shuttle is in orbit, and you can coast around just about wherever you want to go from there. The plan was to use the orbiter like a pickup/delivery truck. Park it with some space lab already in stable orbit, unload the cargo, then bring the truck back home by just letting it fall. I guess if you figured it on miles per gallon it would be cheaper to run than a Pinto.

In the cockpit we were still passing through sloshing and starting to pitch over for orbital entry at an altitude of about 140 miles. With an eerie gradualness, the curvature of the earth appeared darkly at the bottom of my windshield, limned in a faint, glowing light. I was enthralled at the sight. Spaceflight at last!

Distractions pushed the spell aside. The bright moving dot that was us kept trying to escape from the curved lines that was our pathway into the selected orbital window. The more I tried to correct it, the worse it got.

"Fly *to* the needles," the Navy astronaut beside me said softly. I was too dense, or too caught up in the euphoria of spaceflight to translate these meager instructions. The moving bug of light was set up to respond like a flight director does on the instrument panel, and I had almost no experience with this modern and sophisticated flight instrument. I kept treating it like a VOR and got mixed up on how to control the thing. I'm not sure what happened next. Either I barrel-rolled us off into orbit, or stalled the thing in outer space . . .

"Freeze," said the matter-of-fact voice of one of the men seated at the consoles outside.

"That means the computer's got it. Turn it loose." said Navy.

I guess I had just stalled and spun the space shuttle on takeoff. But not to worry. At that altitude it was too high to fall.

The headsets were quiet as the computer picked us up and moved us from the Cape to the desert at 45,000 feet for a powerless approach into Edwards Air Force Base. The desert was tan in my windshield, the low purple mountains were coming up fast. The situation green-screen on the panel now showed Edwards as one dot and us as a moving dot coming toward it.

No one had discussed my problem of flying the flight director backwards yet, so I could not force our long fall any closer to Edwards. With star sensors, inertial measurement system, Tacan, and microwave, I got us lost over the desert from on top of Edwards.

Sweat flowed freely in the very cool, air-conditioned flight deck of the shuttle. The conversation between Control and me sounded just like any other out-of-gas aircraft looking for the airport.

"You see it out your side window yet?"

"Nope."

"Bank a little more over this way."

"Still nothing."

"Well, you are coming on down; looks like you are in for a desert landing."

By this time I had learned to read some of the vertical tapes that show airspeed and altimeter. The horizon showed when I flared to land in the sand. It was awful. The only thing to make it more real would be if they had an engineer standing outside to throw a bucketful of sand into the air intakes and into our ears as the thing skidded to a rocky stop.

The Navy guy gave me a funny look as he got up and left the flight deck. One of the engineers who had been watching this shameful display on an outside monitor climbed into the left seat. He knew me.

"Don't feel bad, Bax. Jules Bergman crashed it 17 times."

He said they still had time to take me back up to 45,000 and try the landing again if I wanted to. I asked him to give me some numbers.

The cement airplane was falling to the earth again. "Okay, you are at 15,000 feet, seven miles out and you are already too close in. Do not let your airspeed decay below 290 knots for approach speed. Remember there is nothing you can do to get any of this energy back."

Edwards was right below us now. No wonder I hadn't seen it before. I was looking for an airport, not a postage stamp.

"We are in the pattern for *that*?"

"We are if you will pitch the nose down more and come on around in about a 45-degree right bank."

I compared the unbelievable view below out the windshield to the display screen. Edwards' runway appeared as a straight line; the pattern was a bright green, ten-mile circle at the downwind end of the runway. The moving dot was still us, but now was preceded by a short string of three other dots. These are called "predictor" dots; the computer predicts where you will be in the next 20, 40, 60 seconds. A great idea for any student training in a Cessna 152. Except the predictor would probably cost as much as a hangar full of 152s.

By forcing the shuttle into a diving right turn, I barely got the leading dot inside the ten-mile circle. Our descent rate was about 15,000 feet per minute. Some glide. The approach lasted about 20 seconds.

"Speed brakes out now, gear down." The thudding, shuddering, and deceleration into the shoulder harness was real enough. I lifted the blunt nose of the shuttle, and there was the three miles of desert runway, right where it was supposed to be, flattening out, rushing towards us. One wing dropped. I picked it up, the other side went down. Oh lordy, we were going to touch wing-down at 180 knots and smear this tile-plated glider down 15,000 feet of paved runway.

"He's entering PIO. Freeze," came the crisp voice in my headset.

I turned it loose and the wings smacked as level as if Bob Hoover was hiding in there somewhere. I gently eased the nose a bit higher.

"Touchdown! Steer with your feet. Brakes!"

The on-rushing runway markers slowed, stopped. I felt wildly exhilarated. Shouted out loud. Wanted to pound on the dashboard.

"Pretty good landing. Bax."

"What's PIO, some terrorist organization?"

"No. It's Pilot Induced Oscillation. The shuttle is so heavy, so short winged, they all do it at first."

"Do they spend much time in here, the crews?"

"Four or five hours a day, five days a week for about 10 months. They get their milk and mail delivered to the doorstep. Their wives write to them here."

"They've really got to be good to land an airplane this hot and know they've got no engines for a missed approach."

"Oh no, the computer can take over and get a perfect landing every time, but none of the pilots want to do it like that."

Me neither. Oh, the fun of flying that slick river rock. It moves heavily, like a DC-3 with the bottom stove out. And the smack of history for just having been there and sat where they sat. Today's kids will look up from Mork and Mindy and say, "What's a shuttle?" A hundred years hence they will show the yellowed pages, "My great grandfather was there when they were learning to fly, even before Atlantis or anything had been built in outer space . . . ."

6

Thigpen and the Landing Gear Handle

THE GREAT IRON LEVER of the manual landing gear on the old-model Mooneys humiliated Ben Thigpen on takeoff, but then I knew it was going to. You can't tell a man that he must grasp the knurled upper section of the landing-gear operation bar as it stands locked into the crossbar at the bottom of the instrument panel, and once flight has been achieved, push the lever forward snug in its latch, grasp the knurled section of the lever firmly with his right hand, unlock the lever by sliding the knurled section down on the lever about an inch or so, freeing the lever from its latch (the wheels now dangling below), and pull the lever back towards its latch located flat on the floor between the seats.

As he pulls the lever down, the wheels now coming up, he must deftly rotate his hand over onto the top of the lever so he can push it down the rest of the way—push it firmly down until the click tells him the wheels are now retracted and the green light on the panel goes out and the red one comes on. All the while flying with his left hand, which requires the lightest touch on the sensitive pitch forces of the airplane, while putting great force on the lever in his right hand.

No, there is no way you can tell a man all that, especially when his first try at it is going to tell him everything he needs to know in a manner he will never forget. After a while, it gets to be as natural to a Mooney pilot as closing the outhouse door while not dropping the catalog, but you could sell tickets to his first few tries.

There must be a point in the gear cycling when there is over 105 pounds of force on the lever. That's what my wife weighs, and when she first tried it, she got her hand rotated over okay. But then she tried to stiff-arm the lever on down, and the increasing force of the plane gathering speed caused the lever to fight back and raised her up off the seat by way of her locked elbow and stiff arm. Her seatbelt was only loosely fastened and there was enough slack in it to raise her, shouting, to the cabin roof, while in the right seat I was laid back and laughing at the struggle. Nothing bad happens if you just turn the handle loose. The gear dangles down half in, half out, and there is lots of drag, but nothing bad happens.

The worst is transferring some of the forces in the right arm to the left and the Mooney pitches up alarmingly, which is what we were now doing. "If you don't shut up that laughing and help me, I'm going to loop this thing on takeoff!" she cried indignantly.

I helped her shove the lever down and lock it, but she had little to say to me the rest of that trip.

Al Mooney was still living then, and at one time I asked him about this, and he said, "There is a window of exactly 85 miles an hour at which the gear can be cycled without any slipstream forces on it at all." The trouble with that is the Mooney on takeoff goes through 85 mph so quickly.

I began to use another system that I learned from pilot talk at one of the MAPA (Mooney Airplane Pilots Association) homecomings at Kerrville. There is no harm in just leaving the gear down until you have the takeoff climb established, the power set, and have gotten up to about 400 feet. At that point, unlatch the gear lever, dip the nose at the same time you swing the lever down, and the airplane flies down over the wheels without effort.

That's why a lot of the old Mooneys looked like they were nodding goodbye to you from the ground. Same thing in reverse on landing. On the downwind leg, reach down and unlatch the gear from the floor by shoving the knurled knob forward, lift the nose slightly, and the handle comes up by itself as the airplane flies away from its wheels. Neat, eh?

Many of us prefer this old "Johnson bar" manual gear for its everlasting and foolproof performance.

The name "Johnson bar" is an obsolete term having to do with the operation of steam locomotives. The engineer had a throttle slightly above his left arm that controlled the amount of steam into the pistons. He had a small brake handle right before him, mounted on the airbrake line, operated in a circular or swiping motion. Rising up out of the floor of the cab right in front of him was a long iron lever that he shoved forward or pulled back to set the valve cut-off with. A wide-open setting would get the long power impulse to start the train, and he would cut it down finer and finer for the staccato chuff-chuffs when he was really high-balling it. Some old-time railroaders claim they could tell who was the engineer by the way the locomotive sounded.

The lever on the landing gear of the old Mooney actually lowered or raised all three wheels in a most delicately designed system of over-center levers that locked the gear in its up or down position. But the power source was all "armstrong."

None have ever been known to fail, except for one that had a "crystalized" weld at the base of the lever. It came off in the pilot's hand one day on final. Troubleshooting was easy, the choice was nil. Ace McCool, who must have been the pilot, shut off the engine and bumped the prop to horizontal with the starter. Most of the damage was to his paint.

"They don't," as you have often heard, "build them like that anymore."

In this departure with Thigpen, a strong broth of a boy, he overpowered the Johnson bar shortly after takeoff, wiped his brow, and gave me a long look. We flew on in peace; our destination was the 1981 MAPA homecoming at Kerrville.

The weather began to close in on us below, near Austin, so we came down from our speed-making 8,000 feet. Thigpen managed a well-executed spiral down through a hole, and we approached Austin beneath a newly formed 2,500-foot overcast. We contacted Austin Approach and stayed in the hands of radar. Austin was murky. At one point, a pair of ghostly blue F-4 Phantoms crossed ahead of us, so Thigpen turned on our landing light and strobes, too.

About 20 miles west of Austin they released us, and Thigpen reached down to the row of look-alike switches on the bottom edge of the panel and turned out the landing light.

At the same time I noted we were two hours into the left fuel tank, and telling him what I was doing, I reached across to the same row of switches and turned on the booster, then to the floor selector and switched to the right fuel tank. All that went smoothly.

The hill country began to rise beneath our swift wings, and San Antonio reported Kerrville about to go IFR in light rain.

Having had one hair-raising approach to the NDB at Kerrville (which sits in a stone-faced valley), and not wanting to try that again, I suggested we just ridge-run as far as we could. My standards for ridge-running, which I learned from an old Arkansas ridge-runner, were to cross one smokey ridge at a time, and only if we could clearly see to the next one.

Just about that time I noticed that all the engine instruments on my side of the panel had gone belly-up. At the same time, Thigpen turned and said the avionics had gone cold too. He suggested that I now fly the plane, since I knew where we were, and he climb in the back and get us a sectional chart out of his bag while we were changing seats.

Our little speedster only bucked mildly as this stoutly built lad wriggled his way far enough back to reach into the baggage bin through the narrow passage between seat back and roof. I settled happily back in the left seat again, being more at home with this kind of flying than I should admit. I checked the circuit breakers on the far right side. None popped, but power was still off. Yet I found enough flight instruments before me to be content with.

Thigpen, chart in hand, was now about to try to climb over the back of the front seat again. He got one foot into the narrow slot between the seats and put his weight on it when suddenly his right knee began to rise and fall and he began shouting something unintelligible with his head pressed to the cabin roof. At the moment, I was dealing with another fast-approaching ridge where the deer and the antelope play, and I couldn't look at Thigpen, but his knee kept rising and falling there beside my right elbow.

One of his size-12 jogging shoes had caught its gripping tread on the handle of the Johnson bar and unlatched it from the floor. Thigpen, now forking the seatback, was engaged in a heroic struggle, a one-legged battle with our 164-mph slipstream, popping the wheels in and out of their wells. The scene inside the cabin looked like something out of an old Laurel and Hardy movie. I'd had too much coffee and was on the verge of the peeing giggles; he was shouting, "Do something!"

I bobbed the nose in a snappy maneuver which settled the airplane back down over her own wheels and brought a high, thin scream from my partner, who had the seat back up between his legs. He got his foot off the gear handle, sort of poured himself down into the right seat, and handed me the chart—wadded and crunched from one of the times his hands had closed spastically around it.

We arrived at Kerrville without further adventures. I made the usual shutdown procedure, but when I came to the master switch I found it was already off.

"Who turned off the master?" I asked. If either of us knows, he has never told up to now.

7

NBAA Convention

I GOT INVITED to the NBAA (National Business Aircraft Association) convention where I fitted in like Bob Wills and his fiddle at Carnegie Hall. The NBAA is just like any other airplane convention, only they don't fly as much or have as much fun as folks do at the EAA convention at Oshkosh.

NBAA pilots don't dare have as much fun, not with expensive airplanes belonging to somebody else. If you come out of a bizjet grinning, somebody will report you to the stockholders, sure as heck.

There were just as many stunning blondes at the NBAA convention at Anaheim, California, as there are at Oshkosh, but the difference is at NBAA they were hired models, at OSH the pilots bring their own.

The distinguishing characteristic of an NBAA conventioneer is his three-piece grey suit and clean shirt. He also stays on the phone a lot. At OSH they have waiting lines too—waiting for the porta-john. Whatever relieves you, as my old uncle Philbert used to say.

At NBAA they have indoor toilets, but no outdoor camping. At OSH you can walk to the flight line; at NBAA they had the convention in a big hotel and convention hall but you had to get on a bus to ride across town out to the airport to see the airplanes on the flight line. The multi-million-dollar airplanes were chocked at Long Beach Airport. Actually you can't tell one town from another in this part of Southern California, it's all paved. Some of the big aircraft companies ran private buses with built-in bars—the only buses I ever saw where they load you on and when you get off you're loaded.

The architecture of the Anaheim convention hall pretty well reflects the local manner of living which is, "If it feels good, do it."

I made a spot survey of what kind of exhibit attracted the biggest crowds. Some of the exhibits were elaborate, more than one story tall; others had the fuselage of the plane they were selling for you to walk through. But the crowds were in front of any kind of a live pitchman. Old medicine-show-style barkers. Big crowds gathered to watch them demonstrate how good their polish was on aluminum or plexiglass. Another big crowd stood before the booth of Jim Bath & Associates, a most conservative firm of international brokers in turbine aircraft, whose motto is, "A reputation for professionalism." They had a card shark there who kept the crowds laughing all day.

Business was bullish, a record attendance of over 12,000, which would hardly be one good morning at the EAA show at OSH. Saw an old friend, a once down-at the-heels charter pilot, and asked what he was doing now.

"Selling Cheyennes, about one a month," he replied, looking fat and sleek now as he snapped his copy of the Wall Street Journal at me.

There was a capacity crowd to hear FAA Administrator J. Lynn Helms. Helms was welcomed by a dense thicket of PATCO pickets who made the mistake of looking like castaways. Sandals and cut-off shorts, beards and hairy bellys. (You mean to tell me *those* guys directed our airliners?)

Helms, who spoke on the record but accepted no questions from the press, said of the unsightly PATCO crowd he had to thread through, "Twenty-four months from now you won't even recall this strike."

That would have forecast it up to September 1983, and he was almost right. He also said, "Aviation's future is not in the past. We will let the pilot do what *he* wants to do, not make the pilot do what *we* want him to." There were cheers, but Helms' crystal ball was a little cloudy on that one.

Assistant FAA Administrator Don Segner had to carry the coals on the subject of airport noise. The reduction of airport noise was a big issue then, as now. I have given the matter much thought and here is my suggestion to help: modify jet engine exhausts to sound like dogs barking in the night. The suburbanites, accustomed to this sound, would hardly notice them anymore.

The NBAA had serious, but dull, seminars in meeting rooms all day, every day. I wanted out.

Out on the flight line at Long Beach airport, the best of the ramp displays was a French Falcon jet that had been flown over "in the green." Actually, flat grey, but that's the term they use to describe an airplane with an unfinished interior and exterior. The French did this to show their engineering and structural concept, inside and out. I went over that Falcon with my fingertips, fascinated with the smooth, butt-jointed, compound curved skin, fastened with flawless countersunk rivets. It was a few million dollars worth of utterly perfect metalwork.

In my mind I contrasted it to the Wichita-built mockups back in the convention hall. Although only shell fuselages, intended to show off elegant upholstery, the ripples, rough rivets, and crooked puttied-over seams of the mockups would have flunked Oshkosh.

I was ashamed of us. But that was back in 1981. Maybe we've learned by now.

The bravest souls were the little contingent from Kerrville, showing off the new Mooney. When you think of executive aircraft, you think of twins, jets, multiengine stuff. There sat the little Mooney, announcing that it was a business aircraft, too. If there was another single-engine, piston-powered plane at the show, I didn't see it.

The NBAA provided chance encounters with some of my heroes, including the tall, distinguished-looking Ed King of King Avionics. I told him all of my six-year-old Silver Crown stuff was still working well in a 13-year-old airplane. He looked like that meant a lot to him.

I listened long to the enthusiastic Jerry Dietrick who is still trying to bring back the all-composite-construction Windecker Eagle airplane. Only about six of these were ever made, and I can't shake the feeling that someday the Smithsonian will be looking for one of these slick, beautiful, non-metal airplanes to tell the story of where the revolution in airframes began.

Although aviation must look toward the future, the most unsettling trend I saw at NBAA was the rush toward the futuristic all-electronic airplane. More LEDs, CRTs, and computer-fed avionics, and bigger and better batteries. One maker was showing a $60,000 accessory that takes the input from your flight plan, weather data, fuel system, radar display, climb curve, cruise range, descent or holding pattern, alternate airport possibilities, and whether or not you want max speed or max economy on the flight. The computer then sets "throttle trim" for you.

I used to know an old pilot who could do the same thing.

But then, old pilots have always had this atavistic outlook. The first Boeing airliners, the M-80s, were made with open cockpits because the pilots wanted them that way. The Model 80-A, introduced shortly thereafter, had a roof over the cockpit—the pilots, now inside, grudgingly looking out the windows.

Come to think of it, my own helmet and goggles are still hanging right over there on the wall. But that NBAA got my head so turned around I nearly forgot about them for awhile.

8

Blup in the Night

BACK IN 1975, the folks in Victoria, Texas, had invited us to speak at their flying club banquet. It went well, and now these new friends were seeing us off from their airport. It was about midnight.

Victoria Regional is one of those huge airports left over from the war that still spatter Texas like giant fallen stars. Nine-thousand-foot runways and no control tower.

There is a special mood about seeing friends off in a small airplane under midnight stars. Victoria is down near the Gulf of Mexico, and the onshore breeze was pressing the ladies' after-five dresses against them and softly rattling distant hangar doors. I was making my preflight inspection of our Mooney.

The ladies stood respectfully back from this ritual, making small talk to my Diane. What could they say? "Sure hope this thing don't kill you." Or, "You couldn't get me up in that little airplane tonight for all the tea in China." No, there was light laughter carried along with the billowing of chiffon skirts that floated and flared in the night air. The men stood to one side, still in evening dress, pilots all; they knew to leave me alone during this preflight ceremony.

I would not hurry the inspection—went at it by rote, as if a fed was standing right beside me. Then it was time to board. Everyone clustered up by the wing root: handshakes, "y'all come back." Soon we were sealed up inside our aluminum shell, a different world with its different demands.

We had only owned the Mooney seven months then. I gave the red glow of the cockpit all of my attention. I was uncomfortable at first with how high the cowling seemed to be. That feeling has long since gone away, but back then it seemed like pulling a Jaguar roadster on like a sock and laying back in it, eyes level with the top of the cowling.

With typical owner vanity I was hoping for a clean start. The whine of the booster, "clear" out the storm window—the starter engaged. One or two blades jerked by stiffly, and the old Lycoming 180 fully awoke. With the sound and feel of power, I felt at home in the airplane.

That particular hard-edged, everlasting Lycoming was one of the main reasons for choosing this airplane. When we bought it, it only had 700 hours total time. A hangar queen. One of its past owners was afraid of it; the other owner had children who got airsick in it. The seven-year airplane was like new, but suffered the usual problems of having sat too long in an open-fronted T-hangar on the Gulf Coast. That's the hardest use she could have had.

Part by part, and in a gentlemanly manner, the Lycoming had let us know what had rotted, what had rusted, what had to be replaced. A little top end work, a rusted jug, all that work was behind us now. The engine had flown several hundred recent hours sweetly singing its song.

We taxied for what seemed like miles across Victoria's airport. There was 9,000 feet of runway available, and I intended to start at the beginning of it. The most useless thing in the world is runway behind you.

We bobbed along on the stiff rubber biscuits of the Mooney gear, the landing-light beam lifting sharp-winged nighthawks off the cement and into curved flight ahead of us. I could sense that Diane beside me was relaxed. Eight thousand hours of Braniff stewardess cabin time. Always marry an airline stewardess. No matter what happens, she's seen worse.

A good runup, checklist in hand. Even that funky right mag was trying to be as sweet as her sister tonight. I cycled the prop twice, always do when the oil is still cold, and carefully set the gyro compass to the runway heading, intending to go on instrument transition during takeoff rather than point the nose up into a sky with no horizon and have to play catch-up. The broad cement runway seemed to stretch ahead into infinity. Check doors, window, belts, "You ready?"

"Ready when you are."

"It's 12:30 A.M. and rolling."

I snapped off the landing light at the first sign of lift—that little pitch-up that the Mooney's flattened laminar-flow wing gives you when lift suddenly develops. The instrument lights had already been turned down to their dullest red glow; my eyes were adjusted to the dark. The transition to flight and full IFR went as smoothly as a drink of cool water.

I intentionally left the gear down until the climb was established, and was just thinking of sucking up the wheels when the engine went "BLUP."

Just like that.

One blup from the engine.

Not a falter in power, every needle standing steady or in the green. Every hair on my body standing, too.

It's funny, all the junk you can think of in such a split second. The words of McKey, who taught me night flying many years ago, came back. "If the engine quits at night, establish a normal glide. At 50 feet, turn on the landing light. If you don't like what you see, turn it off." I almost laughed out loud remembering that. It's true you know.

At 400 feet I cycled the gear up and began a shallow, climbing turn. I wasn't about to touch any of the engine control knobs just then, so our climb was brisk. I noticed I was also keeping us within easy glide-back to the airport, an instinct, like a child clinging to Mother's skirt.

About a thousand feet, easy off the instruments now. There were distant cities, highways, a clearly defined horizon. Beside me Diane had not uttered a sound. Side-eyed, I could see her fine profile in the panel-light glow, but I didn't turn toward her.

I reduced power to cruise climb and plotted locations and distances that would carry us over airports on our way toward home. Galveston, the nearest, 10 nm northeast. The engine never blupped again. Not then, nor ever since.

Close to the Galveston VOR, the runway lights of Scholes Field within an easy glide, I sort of sighed and, for the first time, put my full weight down in the seat. I heard Diane sigh and settle back too. Until then neither of us had moved nor said a word.

I turned toward her. Maybe she hadn't heard it. "You gonna mention it?" I asked.

She turned with a kind of a little grin, her eyes dancing. "Not unless you do."

We both started laughing. I told her she was good troops. She said she knew it.

9

The Texas Gunner

THE GRUMMAN TBF TORPEDO BOMBER is a huge slab of an airplane, the largest the Navy ever flew from carrier decks during World War II. With its 50-foot wings folded flat against its sides, a man could walk under the belly between the wheels hardly ducking. The torpedo bombers that survived the war were often filled with water and used to put out forest fires.

The Navy called them Avengers when they flew at the Battle of Midway to avenge Pearl Harbor. They were expendable in battle, thrown at the enemy fleet. Hundreds of Avengers splashed down into the sea, and hundreds of young men in their prime—young men from Princeton and from the sunny farms of Iowa—died in them, their song unsung, their children unborn. But Avengers were credited with sinking two Japanese battleships, a bargain in the exchange of war.

A flight of six were carrier-launched at the Battle of Midway. One returned. One battleship was sunk. It had been another bargain day at Midway.

In one of these torpedo bombers, the Texas Gunner lay dying. Dying in the belly of that big blue flying whale. With Zeros on its tail, lacing it with machine-gun and cannon fire, and the target ahead—a big Japanese battleship pumping up a forest of sea-water, white columns from point-blank cannon range—this bomber plodded on, as though drawn by the fascination of its own death.

The thick-set shoulders of the New York pilot never flinched. His smooth, faintly olive-skinned face was impassive with the concentration of holding his nine-ton craft low and steady over blue water. To aim his deadly drop of TNT torpedo fish straight and true, he knew he must aim his body. He sat behind a racketing 14-cylinder Wright Cyclone engine, which delivered a steady 1,600 horsepower and blew a gallon of oil an hour across his wings and windshield. A good engine. We used them in B-25s, A-20s, and Mariner flying boats during the war. With a gauntleted hand the pilot cracked his canopy back, admitting some cool air and a new shriek to the sounds of war.

Beneath the long, greenhouse canopy behind the pilot sat a tall blonde Aussie who was swiveling his head, the sun gleaming off his balding pate. He was excitedly pointing out new targets to the Texas Gunner, who was in the turret just behind him. The Aussie jabbed a long finger into the sun where a bent-wing Corsair was dropping into a death dive onto the Zeros on their tail. Did the Texas Gunner see that? That beautiful fighter. "Whistling Death," the Japanese called the Corsair. Now they had a chance after all. And look, a tubby Navy Wildcat and a sleeker Hellcat had joined in pursuit of the Zeros. A wild melee of fighter planes swirled soundlessly around the torpedo plane. Did the Texas Gunner see all that? Breathtaking!

But the Texas Gunner was no longer in his turret. His lean warrior face was now green, his blondish-red hair now matted, he lay a-dying deep in the pit, back at the narrow ventral end of the echoing, roaring, aluminum cave at the rear of the TBF.

No more Alamo, no more fierce screaming rebel yells; the Texas Gunner lay sprawled over the hump of the tailwheel housing, staring up at the ribs and rivets of the Grumman Iron Works tailcone.

With one deathly white taloned hand he hung on for life, while his fast-dimming mind was thinking, "Quickly, Lord. Oh, let this be over with quickly. Let those long tearing slashes of the Zero's machine gun bullets open this aluminum and find me. Let us collide full-on with a white column of sea water from that battleship cannon fire ahead." He would even accept the worst way to die in a TBF: if the pilot got hit and they would all have to ride it down with him. Only the pilot had controls in this airplane. The crew could see them, but there was no way to reach them.

He stared forward at the rugged ridges of the torpedo bay bulkhead and its cluster of hydraulic pumps and thought of how it would shred him if a wounded pilot managed to fight his way back and crash-land the bomber on the carrier deck.

The Texas Gunner groveled in the bilges of the tailcone as the Avenger swung back-and-forth and up-and-down in the gusts of hot, oily air that found its way back there. The Texas Gunner wasn't dead yet. Not even wounded. Or scared. He was airsick.

With the honor of the Lone Star state to bear, he lay limp as a boiled green string bean, down in the hole, missing the whole spectacle of the Battle of Midway.

Born and bred to be a hero, all his life dreaming of being in real air-to-air combat, seeing fighter planes in action—and now look at him. "I will not barf," he resolved with all the courage he had left in him. "And if I do, I will use my own shoe. They must never know."

A Zero flashed by, red meatball on the wings right over the canopy. "I've got to see this." He began to crawl, hand over hand, up the inner works of the deep fuselage, snatching his hand back from a mistaken grip on the control cables just as the pilot hauled up in a victory zoom over the Japanese battleship blossoming into flames below.

The Texas Gunner managed to drag himself upwards until his eyes were level with the turret sill just as the pilot of the Avenger sucked the big airplane around in a roaring vertical-banked turn. With nose squashed against the turret ring and knees buckling under the G-loads, the Texas Gunner got one glimpse of the flashing wings of a genuine fighter-plane dogfight, then sank with a gurgle back down into the dark bilges below, missing the canopy-to-canopy victory roll of their beautiful escorting Corsair.

He hugged the tailwheel housing again. "Dear God, if I can only hold it until I hear this tailwheel wham into the runway and roll out, I will never sin again."

Fifty thousand cheering fans stood up in the grandstands at Rebel Field, Harlingen, Texas, Summer 1979, as the Avenger and all the rest of the Confederate Air Force taxied by and the brass band played *Dixie*. Even the Zeroes were part of this heroic parade. The pilot of the torpedo plane and the Aussie were waving back to all the beautiful girls who were jumping up and down and applauding in their midriffs and shorts. The Texas Gunner stayed down in his bilge, peeking out a tiny window. "If I couldn't stomach seeing the Battle of Midway being replayed with its original cast, I'm sure not gonna be hypocrite enough to stand up and gather roses now."

And I stared all around at the gaunt, green-ribbed cavern of that old Grumman and tried to imagine the courage of the 18- and 20-year-old kids who once did this for real. No way. No way could I envision that much courage.

10

Approach Wants You on the Phone

IN THE AUTUMN OF 1979 I met an old friend at the airport coffee shop. Even holding his cup with both hands, it still clattered when he set it down on the table top. His eyes still held some of the fear of whatever it was he had just seen.

"You want to tell me about it?" I asked, easing in across the table from him with my own cup of coffee.

"Yes, but only if you write it so that the names and places could never be traced, because I know you, and I know this is a story for the magazine."

So none of the names or numbers in this story are real except the common numbers having to do with navigation. The story is how Claymor Klutz, an experienced lightplane pilot, nearly got into a midair with a Boeing 727 while both were on approach and both were under radar control and both were talking to Approach Control and it was still daylight VFR. It would have been another of those mysterious tragedies the press would have had a field day with.

Klutz is a commercial pilot, instrument- and multi-rated. Statistically, he qualifies for and gets the best insurance rates. Personally, he still likes flying, though he's been a pilot for years. He respects the art and wants to be considered good at it. Not a whacko.

Today, Klutz had departed homeplate in the sunshine, flying his own high-performance single-engined plane on the short and very familiar trip over to the big city international airport we shall call JET. A passenger pickup. The sort of thing he often did.

He was familiar with JET and knew he'd probably be the only small aircraft in a long string of inbound turbine-powered transports. But he was "up for it" and had laid out his Jepp plates and a notepad upon which he had copied all the anticipated frequencies in the order of their expected use. He did not, however, have a diagram of the TCA in front of him, intending to stay low and report a known visual fix ten miles out: the lake.

He called in at the lake, "JET Approach, November 1234 reporting the lake, 1,000 feet, squawking 12, have Information Kilo, landing JET."

Approach came back smartly, gave him a squawk code and a heading of 270 degrees, and told him to stay clear of the TCA and to contact Approach on a different frequency. Claymor rogered that; it was what he expected. He had the new frequency already written down, and he set his new squawk code. The only thing he was wondering about was how could he fly west and stay out of the TCA when the TCA was due west of him? Surely Approach knew this too. The lake he had called-in over was the only lake and was east of the big airport. Oh well, he plunged on at 100 knots wondering, "What next?"

Next, he noticed lowering ceiling and reduced visibility. Claymor decided to slow down, descend, and contact Approach about this. This second approach frequency, which he had been monitoring, did not carry the airline traffic, and he missed not being able to form a picture in his mind of where the traffic was and what it was doing.

He had to call this strangely quiet Approach twice, and all they did was hand him off to yet a third Approach. He had to write that one down; it was not in his notes. He was pretty sure that by now he had punched into the side of that big birthday cake, the TCA that he'd been told to both keep clear of and fly toward. Crazy man, crazy.

He descended to 800 feet to maintain VFR, knowing the countryside and knowing there were no tall surprises out there. ATIS for JET was reporting five miles. It looked more like two to him.

The third controller was not so quick to come up. His frequency was eerily quiet for all the traffic he knew must be out there. He told the lethargic voice of the third controller that he was descending to

500 to maintain VFR. "Roger, you are two and one-half miles from Runway 27, report field in sight."

Claymor decided they must have known what they were doing, so he continued on, low and slow. That transmission also confirmed to him that he was deep in the TCA and under radar control.

Just then the sleepy-voiced third controller told him to start a wide 360-degree turn to his left. "Bumped off the localizer for faster traffic," he thought, again wishing he was in on the busy circuit so he could hear what was coming up behind him.

The third controller sounded calm, even considerate. "Caution 1234, you have a water tank on your present heading." Claymor was familiar with the low tank, reported he had it in sight. Then Approach asked in a most everyday voice, "Do you have a 727 on final at your nine o'clock, about one mile?"

Good grief! Claymor reported no contact.

"Straighten your circle, fly north, 360 degrees."

Now Klutz was sure they were threading some faster traffic through. They gave him the tower frequency, tower cleared him for an immediate landing and asked for a short turnoff. Claymor popped gear and flaps, landed on the numbers, and easily made the high-speed first turnoff. Tower sounded aggravated, gave him the Ground Control frequency and a land-line phone number, and said, "Approach wants you to call them." Right there within earshot of everybody. Klutz got rattled trying to sort out phone numbers and frequencies and was humiliated to have to respond, "Say again?" The tower gave him the wrong number for Approach anyway.

He taxied to the transient ramp. The line boy came out, looked him over, and said "Are you November 1234?"

"That I am."

"Approach wants you to call them."

"So I've heard."

Klutz tried the wrong number from the fuel desk and had to call the FSS to ask for the proper number. They didn't want to give it to him until they were convinced that Approach was really on his case about something.

"All this time I was leaning over the fuel desk, and the girl and the line boys were listening and watching me sweat."

He finally got Approach on the phone and identified himself, and they got right on his case at once.

"According to our airport surveillance radar we had you inside the TCA without a clearance. Once, we had you right on the glidepath of incoming traffic. Are you aware of that?"

Klutz was not sure of what all had happened, or who had talked to whom, but he remembered an old east Texas saying: "Lots of people start fights in the police station, but the police win 'em all." He decided to be as nice as he could be. He said, "Yessir, I'm aware of it now, and I sure do want to thank you folks for taking the time to discuss it with me."

"My mind was doing a fast re-play of all that had happened in the last five minutes, and I was pretty sure one controller was not aware that the other had vectored me into the TCA. I didn't really belong to anybody on a scope until I showed up doing that 360 across the glidepath. Then I was everybody's meat."

Claymor still looked miserable about it. He continued. "I wouldn't intentionally bust a TCA, and I have no respect for dumb bunnies who do, stumbling around out there on the approaches of fast traffic. My God." He shuddered at the memory and the thought of what all could have happened.

"What'd they do to you?" I asked.

"Nothing. I took the rap for it, and all I got was the one phone call and the contempt in their voices."

We sat there in silence awhile.

Then he said, "This thing had all the makings of another private-plane/jetliner midair. If you had been one of the investigators, what would you have told the press after you had listened to the tapes? Who would you have tarred as the bad guy, the killer?"

I asked him to let me think about it awhile. We sat in silence. He really looked miserable, and I don't blame him. Then I told him it looked to me like Approach had too many frequencies and those controllers were not in close enough touch with each other about what each was doing.

He said, "I'll roger that."

The thing was over, and nothing bad had happened. It wasn't even a genuine "near miss," because he was being vectored the whole time. But none of us ever forgot about it.

11

When the Tri-Pacer Was New

THERE WAS A TIME, back in the long delicious summer of 1956, when we would sit in the shade of the airport cafe, pick at blades of grass, and seriously argue about whether the Tri-Pacer, which was new, was a better airplane than the Cessna 172, which was also new.

There was much to be said. Some believed the Tri-Pacer was faster, its 150 Lycoming stronger by five horsepower than the Cessna's mild Continental. And nobody could argue that the Tri-Pacer wasn't quicker, more nimble, more fun to fly.

The Cessna 172 guys kept harping about all-metal, but most of the airplanes around us then were fabric. Some of them were surplus from the war, and they were all still pretty good.

It was a kind of a Ford-vs.-Chevy argument, based more on whether you liked Pipers or Cessnas rather than fabric or metal. And back then more of us had learned in Pipers than in Cessnas, and there were more Pipers around, parked where Cessna 150s soon would be.

Nobody mentioned that the Piper had funny little squirrelly ways. Most of the planes we flew then were a little squirrelly one way or the other. The load-hauling difference between the Cessna and the Piper didn't seem to make much difference either. They were both four-place airplanes. Weren't they?

Gerald Ashley, the local Piper dealer then, had his brand-new Tri-Pacer parked right out there on the ramp, rib-tight and gleaming in the sun. He heard me coming down on the side of the Tri-Pacer and asked if I'd like to get checked out in his. Hell, yes!

The airplane was dark red with black trim and had fine black fabric upholstery inside. A nubby-weave cloth, with red piping. It was a luxurious airplane. Smelled like a new Cadillac, only better because of the lingering whiffs of fresh dope and gasoline. And there was a rear door for dignified entry into the rear seat. Take that, you 172 guys.

Everything about this airplane had the fine fitting, attention to detail, and good, solid hardware that you would expect from something that cost over $4,000 new.

It had little plastic cups, set into the windows, that you could turn by hand to get real blasts of fresh air. The master switch and the starter were hidden away under the seat to discourage theft. The fuel-selector switch was hidden away, too. Anybody who took the plane and didn't know all this was not going to get very far. Gerald claimed you couldn't stall it or steal it. It even had a radio—came from the factory with one. You could crouch down, peer into it, and turn a little crank. Lord knows how many frequencies were on it. Twenty-five or 50—more than a man would ever use.

Standing back and looking at the airplane, I saw the Tri-Pacer as the natural outgrowth of the Cub, a Cub with a wider pelvis and turned-around landing gear. Its hip bones stood out rather sharply at the end of the cabin area where it tapered quickly down to a real Cub tail assembly, wire braces and all. Under the wing it had the same strong-looking V-struts. Like a Cub, it had fat tires and big bungee cords, wrapped around a split axle, to soak up whatever may come. I was warned to never let the nosewheel touch down first—but who, for crying out loud, would land an airplane like that? It was an altogether stout and trustworthy-looking airplane.

I had a friend, a high school teacher, who still had a Piper Pacer, which he claimed would fly rings around the Tri-Pacer. It did look sharp, all cocked-back on wheel pants like that, but we ignored it. Nosewheels were the coming thing; they made flying so easy. Cessna ads were all about "driving in the sky." And at Lock Haven, Piper claimed they had taught an eight-year-old little girl to land and take off in a new Tri-Pacer in just six hours.

When it was my turn and I was ducking and grabbing for that under-dash brake lever that set both brakes at the same time during a high-speed rollout, I decided that must have been one heck of an eight-year-old little girl.

Ashley spent most of our time aloft showing me that the Tri-Pacer just wouldn't stall or spin. He would hang it on its prop, that little four-banger rhythmically beating at our fabric bass drum, and show me how strong the rudder and aileron control was. The stubby little airplane would just hang there, its short, fat, Cub wings grabbing at lift but never twisting out from under us. I had been flying a Luscombe that, if you treated it this way, would have given you the thrill of your life. Later, when I flew the Tri-Pacer by myself and did the same thing with it, I discovered that, while it was pointing almost straight up and was not trying to spin, it was also falling about a thousand feet per minute.

My old 1956 logbook shows that I had 47 hours total time and that Ashley spent 45 minutes with me before turning me loose with his new Tri-Pacer. I had just gotten my private pilot license the day before, and that was in a 172, which was my first time flying anything bigger than a Cub, Luscombe, or Aeronca. With the ink still wet on my license, I was already into high-powered luxury airplanes.

The following Sunday all the local sportsmen, who usually sat in the shade in the grass and talked, decided we'd do a fly-in for a steak dinner at the H&H Guest Ranch near Houston. Can you imagine that? A good restaurant with an airstrip instead of a parking lot. That would still be a good idea today.

I found three friends who were satisfied that I must be a pilot if I had a license, and they accepted my invitation to go.

It was already hot that morning, but I didn't worry. I filled up the gas tanks and all the seats. It had four seats, didn't it? It sure took its time in taking off though. But I was right up there with the 172s and Stinsons—just as much class.

The next time I noticed anything funny was in making the steep turn to line up for a short landing on that grass strip beside the restaurant. I had a thousand feet when I turned onto base leg but only 500 feet when I rolled out of the turn. I leaned over and tapped the altimeter to see if it was okay.

I was on final now and had those little bitty flaps down, but the part of the view ahead that was neither moving up nor down indicated I was going to land a considerable distance short of the airport. I applied full throttle and began to use up some of that slow-flight lift we had been bragging about during the demo ride. This put the touchdown spot back in the windshield, and I made it to the field, but as soon as I reduced power it landed.

I didn't think much of this, but coming back in the cooler part of the day with two hours of gas gone from the tanks, it took off and landed just like a regular airplane.

Many years later, after we got the Mooney, I was flying the airways and saw a little dot ahead at same altitude. Slowly closing the distance, there grew a little Tri-Pacer, beating along bravely. Closer now, and from a shortened rear-quarter view, it really did look like a flying milk stool.

Alongside him, I throttled back just to look and keep pace with the memory awhile. Very little throttling back, he was fast. We looked each other over; he waved; I waved back. And I remembered when the Tri-Pacer was new. It seemed so much bigger then. Such a good little airplane.

I knew it wasn't Ashley's. In that first one he had soloed a student who couldn't find the fuel selector switch, ran out of gas, caught a ditch in the field he was landing on, and rolled the plane up in a ball. He walked away from it. Good little airplane.

12

Skydiving Cats (Part I)

THIS STORY STILL HAUNTS ME. I still get mail about it. Anywhere I go as speaker, somebody, some idiot, is sure to ask me about it. I may never get away from it. It all began simply enough.

I got this letter about a skydiving cat. The letter was from Robert J. Hamp III, of Roscommon, Michigan. I answer all my mail, but this guy I called on the telephone.

The letter said that up on Bois Blanc Island in the Mackinaw Straits, there is a cat that skydives. "As I understand it, these fellows have a Champ. They carry the cat up to something like 3,000 feet and slide him out through a six-inch-diameter piece of stove pipe. I'm told the cat is waiting to jump again when they taxi back to the grass ramp."

I put in a call to Mr. Hamp. His wife, Merrily, answered the phone and said that Bob was out on a trip. I asked what he flies, and she said, "Nothing right now. He landed his Cessna 140 on the lake yesterday, caught a snow drift, and stood it on its nose." Merrily sounded gleeful telling me all this. She said she would have her husband return my call when he got in.

It's always good to know a couple where the wife takes such a happy interest in her husband's flying.

On the phone the next day, Bob Hamp admitted he now had a souvenir propeller to hang on the wall, and I welcomed him to the club. "But about that skydiving cat?"

Well, from here on, the story begins to deteriorate. Hamp, it seems, had not actually witnessed the skydiving cat, but his impeccable source and only contact with the catdivers out on remote Bois Blanc Island was one Mr. Allen Hoffman.

Hamp described Hoffman as one of the last living Michigan barnstormers, a retired flight instructor, FAA designee, and ex-manager of the Harry Brown Airport at Saginaw.

As well as I can follow all this, Hoffman now lives on Bois Blanc Island, is in his 80s, and flies his old Taylorcraft off a 2,500-foot turf strip, a beautiful place with a deep-water well right beside the airport.

Attempting to steer this fact-finding mission back to the cat, I replied, "The wild north country is full of 80-year-old T-craft pilots, and they sit around the stove all winter and think up lies and drink whiskey. We are not doing a travelogue on the sod strips of the Mackinaw Straits. Tell me about the cat. Is it a tomcat or a minnie cat? A long-haired cat or a short? Does anybody know what the cat actually thinks when it hears an airplane engine start up? I don't want to sound unfriendly, but this is a long-distance call from Texas, and it's costing money. Do you happen to have the names of any of the jump-plane crew of this furry, falling feline?"

Hamp, not the least chastened, reminded me that hangar tales are flying's true line of communication about the actual events that take place within the aviation community, and this would not be the first time that the media is the last to find out what is happening.

Hamp was unable to answer further questions with any hard facts, but he promised to assist us by putting his full resources into a detailed report in time for the deadline of our next issue.

Meanwhile, I have launched a full investigation of my own from Texas. I have a call in to our GADO office in Houston requesting the FAA's position on cat launching and whatever category changes may go into effect when a six-inch piece of stovepipe is installed on a Champ. Is the airplane still utility class? Experimental? Cat 1?

I have already contacted Dr. James A. Massey, DVM, member of the Flying Veterinarians, a pilot, and an aircraft owner. After an initial outburst of high boyish giggles, Dr. Massey began to talk about the terminal velocity of a cat. He mentioned the spatial orientation and tail-steering-while-in-flight characteristics of cats.

He said the spread of skin might be adequate for an average eight-pound cat, but that a long-haired cat might do better than a slick one. Something about "partial vacuums formed in the fur."

At this point, Dr. Massey was seized with another outburst of the giggles, and whatever his technical point was, it was lost to me.

I intended to supplement my information with calls to NASA and to Mobil Oil, the latter having had some early experiences trying to fly a red horse.

An inquiry at the Society For The Prevention Of Cruelty To Animals was brushed off with, "If this cat persists in hanging around small airports and the pilots of light aircraft, then the matter is no longer of any concern to us."

It has never been our policy to run a "to be continued" magazine article, but until all the facts are in (we are now looking for a photographer-skydiver with enough hair to get us some air-to-air shots, or at least some closeups of skid marks inside that piece of stovepipe), or until confirmation from Bob Hamp, I'll have to let this cat tale hang here.

13

Oshkosh Tower

EACH YEAR the staff of *FLYING* goes off to Oshkosh to publish the *EAA DAILY*. Usually about eight pages, stapled, lots of photos, faces, current convention news. This is the only time all of us at *FLYING* ever get to be together and to work together. Along with the regular staff at One Park Avenue in New York City, there are lots of stringers like myself who send in copy from out in the hustings. Except for that wonderful week at Oshkosh, we never even see each other.

We are all crammed into a small metal trailer: artists, editors, writers, photographers. We, who are used to a three-month lead time on any story and a leisurely once-a-month deadline, now have a deadline every day at 3:00 P.M. The closer it gets to 3:00 P.M., the wilder the scene gets inside the trailer. We love it. It's one of my favorite times of the year.

This story is not from *FLYING*, but one I did for the *DAILY*. There were about a half-million people at the 1980 EAA convention. In 1986 the number almost reached a million. At times, Wittman Field was closed, not because the tower could not handle the traffic, which was the world's heaviest—and with a mix of everything from jets to 30-mph homebuilts—but because there was simply not one foot of ground left to park another airplane on.

With a record of never a serious accident, the world's heaviest air traffic flow is managed by a unique FAA system of "uncontrolled control." Pilots on their way to Oshkosh can get a printout from their local FSS and use that as a guide for getting into the traffic flow and what to expect. Mainly, don't expect to talk back to the tower. Some of the tapes of tower control are sold in the EAA store as sheer amusement. Some European controllers come here to see how in the world we do it.

An Oshkosh arrival is a flying experience no pilot is likely to ever forget. Here is how it was on Monday, August 4th, 1980:

There were three talking controllers in the Oshkosh tower, and each of the talkers had another controller at each elbow who did nothing but use powerful field glasses to spot and softly call out incoming traffic.

The spotter gave the color, type, and location (by local landmarks) to the play-by-play talker, who had to be able to talk, listen, and make decisions simultaneously. The talkers spoke close into their mikes in low, steady, controlled voices. Leaving streams of cigarette smoke, dousing parched throats with cold coffee, they all paced with the held-in passion of caged cats.

"Red Tri-Pacer, put it on the numbers . . . Cessna, clear to land beyond the Tri-Pacer, Ultralight, start your turn now . . . Mustang, where are you? Flash me a light."

Mike switch off, the controller said to his spotters, "I don't see that damned Mustang yet."

The skies were as low and grey as skimmed milk. Nobody could find the Mustang.

"Cessna with the light on, real good, do just what that other Cessna is doing. Beech 18, don't you cross that runway."

They had two or three patterns working. NORDOs (planes with no radios) were using the parallel taxiway for another runway. The rumor is true that IFR traffic sometimes gets in faster, but only up to the point when you get into tower control; then you get sorted out and brought in with the herd.

Takeoffs were taking place too. Ignored by the tower controller, takeoffs were flagged off by controllers actually standing out at the runway threshold. By hand and flag signals they brought an airplane to the edge of the runway and signaled him to hold his brakes and run up his engine. Then, when they see a hole in the landing stream, they point the flag at the takeoff pilot and fall to one knee like a carrier launch. Go!

Steve Schimmig, takeoff controller and pilot himself, said, "Sometimes I wish I had a foxhole to dive in. You need lots of tread on your tennis shoes for this."

In the tower, "Where's that Mustang? He's big. We should be able to see him by now . . ."

Then the wind shifted, and they got ready to turn this whole circus around and use a different runway. Ron Hellman began to tape a clear, concise new ATIS for immediate use. The runway switch was being fed to the pilots, too.

"Dark Bonanza, you will be the last to land on Two Seven. Aircraft following Bonanza, come left, right over tower and enter a downwind for Runway Niner. Ultralight, cleared to land on One Eight." The ultralight had just appeared out of the mists, cocked over to the windward, and landed and stopped on the diagonal width of Runway 18. Nobody commented. They were asking each other if anybody had seen that Mustang yet? "He's fast. We ought to have him by now."

There were worried looks.

A gaggle of AT-6s swept over the tower, their big radial engines shaking the glass. Tower left them alone. They were a part of the fly-by pattern.

After each EAA show, the EAA and FAA meet, play back some tapes, and rehash the events. Out of this meeting come the decisions on what changes they need to make. So simple. So effective.

Tension was building in the tower cab about the yet unseen Mustang. A P-51 has big blades on it.

When a controller's voice dries up, he swaps places with a spotter. Sometimes they will go into a ten- or twelve-hour overtime day. Yet Tower Chief Ryan Gove says his only problem is sorting through all the applications from control towers all over the country, from people who volunteer for the Oshkosh job.

Ryan explained, "The fun, the thrill of it, this a highlight in a man's career. And it goes into his permanent file too. We mix a few new ones in with the old hands each year. We invented this, us and the EAA. The rest of the country still doesn't believe it can be done."

Will Washington take note and use the speed and flexibility of the Oshkosh landing system in saturated airports elsewhere? Ryan said he'd rather not risk an "on the record" reply to that one.

But Ken Jackson, a specialist and tech advisor out of the

Minnesota regional office, said some of the FAA high brass flew in themselves in a Cessna Citation jet during a peak time. "They were amazed at what we can do."

Tensions mounted in the cab. Everyone was pacing, looking out there for that Mustang. Jackson declined to comment on the rough language used off-mike. "That's how they blow off tension. Have you ever heard doctors talking to each other during surgery? The language is awful. Same thing."

About that time, the controller expecting the Mustang arrival raised his voice so that all in the cab could hear him. "Did you say you are a *Midget* Mustang?"

He closed his mike and let his hands swing down, rocking with laughter. "Why that little sumbish! He ain't no bigger or faster than the rest of that stuff out there. Oh, that lil' bastid!" The control room rocked with laughter, and the bubble of tension busted. The homebuilt Midget Mustang came scooting in, never knowing what a time he'd given the Oshkosh tower.

14

Eleanor

ONE OF THE BENEFITS of writing for *FLYING* is that people write back. Any letter that praises or takes us to task goes "around the chairs." Every editor reads it, and copies are sent out to everyone mentioned in the letter. The good and the bad, we take them all to heart.

I answer all the mail that is forwarded to me here in Texas, and worse yet, am a letter keeper. A pack rat. In this instance, I'm glad, because I want to share my heart with you fellow fliers and tell you of Eleanor. Her last letter caused me to dig out the files and find all I could about her.

Eleanor Blakely sent me a card for Christmas 1978. It read, "Isn't my bird lovely? We're doing fine and will start cross-country lessons in the spring. This has been, and is, the best thing I ever did . . . very grateful for it. Love, Eleanor."

The front of the card shows Eleanor, a tall, rangy, handsome woman of indeterminate years. She wears a black tailored suit, flowing cravat, short-cut blondish hair. She is standing in the traditional proud pilot's pose by the nose of a most average-looking Cherokee 140, which can be the most average-looking airplane in the world. But not with the touch this lady's hand is giving it. In the background, lettered over the hangar doors, are the words "The Pennsylvania State University." I had marked the envelope "do a Bax Seat on her." That was three years ago.

The next letter I could find from Eleanor was dated June 1980; with it came a clipping and a newspaper story by Robin Meckler, of the Centre Daily Times, June 6th. The photo, taken in winter, shows Eleanor, of State College, and her two "little brothers," two children named Robbie and Billy, ready to go for a hop in her Piper. All smiling and bundled up in the snowy sunshine.

The story begins, "Not all volunteers of the Centre County Youth Service Bureau can take their little brothers or sisters for their first airplane ride."

Then the story details how 60-year-old Mrs. Blakely, writer and editor for Penn State's College of Agriculture, became a volunteer for the "Big Brother" program and uses her own airplane, acquired two years ago, to reach these troubled children.

"Many people can become so involved in academic or domestic life that they fail to realize they're missing a great deal of new experiences until it's too late," said the flying grandmother. "I love my job, don't plan to retire until I'm 70, but when I do, I'm not going to be stuck deciding how to spend my time—and I'm certainly not going to stay home, cook, do laundry, and pick dog hairs off the rug."

In the accompanying letter to me, Eleanor said the little kids were delightful, bright, affectionate, and noisy. When she took them on as their "Big Sister," they were making D's and F's, and one was repeating second grade and lived in a "junkyard." They loved flying.

Then she detailed her anger at finding out that "Ladybug," as she calls the Cherokee, had cancer. She found corrosion problems in November 1979, and four months and $8,337.86 later she got her airplane back. "She's virtually rebuilt, new metal on her belly, spars treated, brake and fuel lines replaced, even new upholstery; she smells like a new airplane." But Eleanor was sizzling angry over mechanics who had passed the airplane during previous annuals and not noticed that it was rotting away.

In her confrontation with the FAA inspector, she tartly told him that if they were as attentive to the DC-10 as they were to her Piper 140, the 1979 Chicago affair may not have happened, "which he didn't like at all. Actually, I knew he was only doing his job, but I thought the point was well taken."

She concluded the June 1980 letter with, "I've been flying anything I can get my mitts on, now current in the Piper 180 and Cessna 150 (and am aiming for anything more than I can handle, like a P-51). Total time now 300 hours, 135 of that solo, so I'm coming right along. Enough time now to get into real trouble, as one friend put it. Or to be considered a real pro, as another said. My best, Eleanor."

In December 1980, Eleanor wrote again. "Many thanks for the kind words. In the flying world, I am regarded as a creature from another planet, doing something wonderful and weird. There sure is magic involved . . . "

"This painting of me and Ladybug aloft was done by a friend. My instructor says I'm obviously flying into IFR weather. I did have her up in the rain last week, just ahead of a storm, and didn't push anything . . . My life in the air is beginning to sound like the *Perils of Pauline* . . . ."

Then Eleanor detailed a letter from the FAA, grounding her for a cardiovascular condition. She was taking a preventive medicine for mild high blood pressure. Eleanor called Audie Davis at the FAA in Oklahoma City, who said the first doc had read her EKG wrong. "You're fine, go ahead and fly."

Eleanor comments, "I'd like to give that first doc rabies by biting him. Audie Davis, I love you."

"Then my mechanic woke up in October, they had not done the engine annual. A dangerously worn item: the cost, $1,476.01. Must say it's wonderful how people seem to enjoy lending me money to fix my bird."

Eleanor tells of talking to the entire third grade of 52 wiggly nine-year-olds, and how Robbie's counselors are delighted at how much he has improved since she took him on as a flying friend. "Billy is doing better, too."

Then on July 20th, 1981, there was a familiar envelope from State College—this letter penned, not typed:

> Dear Bax,
>
> So you think I've vanished. I learned to my astonishment a week ago that I have wide-spread cancer that will kill me in three or four months.
>
> I must say that this does change one's outlook. There are complex and painful treatments that could be tried, but would do no good. My doctor sees no reason to mess me about, nor do I. I'm not grabby for the last twenty minutes.
>
> All I can think about is what a wonderful life I have had

and how grateful I am. The outpouring of love and affection around here has been incredible to me. My dear old instructor finished my biennial this week, not that I'll need it, but he knew it meant a lot to me. No more solo flying, but there are about 20 guys around here glad to go along. I'd have liked another 20 years of living and flying, but had a wonderfully rich life.

My little airplane has been financial insanity from the start, but has been worth every bit of it, and I'm enjoying every day I've got left.

Love,
Eleanor

This 61-year-old flying grandmother signed this one as she did the first one, "Love, Eleanor."

Dear Lady,

I sit here now, tears in my eyes, only four years younger than you and acutely aware that it's autumn for me too, and that there are no exceptions. I am going to do a thing unusual for *FLYING*. I am sending you a copy of this manuscript at the same time I send one to my editors. First, to let you claim privacy if you wish; secondly, because there is a 60-to-90-day lead time in the magazine, and I want to be sure you see this story. I want you to know we shared your joy of flight, and your courage, with the world community of pilots.

Many pilots before you have set a certain style in their manner of going, for that is indeed a person's last choice. We admire your fiesty grumblings and the manner in which you took care of Ladybug. And you have set an example for all of us in the keeping of a tradition that the Wrights would have admired, in the use of aircraft to aid the lot of others.

You may not have been the first to fly across an ocean, or a polar ice cap, or to leave us a romantic poem in the manner of the young fighter pilot who wrote *High Flight*. But you have written this, and earned for yourself and Ladybug a place in these hallowed halls of aviation.

And how we cherish such things. I will ask each pilot, the next time he or she is aloft in a glorious sky-filling sunset, to dip the airplane's wings and whisper your name.

Thanks, Eleanor Blakely.

Love,
Gordon

Proof of the deep human chord that runs among us, I received more mail from Eleanor's story than from anything I have ever done, before or since. Including subdued thanks from her immediate family.

And no, she would have not have lasted out the lead time and seen a copy of the magazine, but for a little miracle that took place at One Park Avenue in the big, cold city of New York.

The magazine came out early. Especially her copy. Nothing was said of this among us, but as soon as a "galley proof" was ready (that's the finished page before it goes to the binders and distributors), a copy of the galley proof, with the cover of the magazine and Eleanor's story set in type, somehow reached her bedside at the end, but in time for her to read it.

To all of you who cared, and care now, for this valiant lady, again, thanks.

bax

15

Merced

WOMEN AND AIRPLANES LAST FOREVER in California. It may be the soft air.

I have never seen so many immaculate Stinson Gull Wings, Wacos, Fairchilds, Boeing biplanes, and go-go grandmothers, sleek in shorts, halters, and tan. So many stunning memories of the thirties, still in casual everyday use.

Oh, I have prowled the ramps back east at Oshkosh, Ottumwa, and Reading, but have never seen such a collection of finely stitched wings as gathers in the golden sun each June at Merced, California, there on the desert sands between mountain ranges so high.

"Why hasn't anybody heard about all this?" I cried out to the sun-seamed Mac Duff, who had sort of been assigned to me.

"Because none of you guys publishing beyond yonder mountains ever write about us." He grinned, waving a copy of Dave Sclair's *WESTERN FLYER* under my nose to show that Merced was no secret beyond the Rockies, and then holding spread a copy of the town newspaper, the *MERCED SUN-STAR*, which, during fly-in week, looks like an aviation publication. I mumbled apologies for all of us, pocketing the papers for further study.

Merced is somewhere out in the desert of northern California, somewhere beyond the valley of the grapes. To get there, you turn right after leaving San Francisco. It's about a two-hour tankful in a

Stearman. I flew my friend Lee's Stearman to get there, him in the front seat.

We were a loose gaggle of four airplanes. Lee's bunny followed close in a Cessna, glaring at the back of my helmet; I occupied the seat she usually sunbathed in. The third airplane was another beautiful Stearman, rising and falling beside us. The fourth was the winged phallus silhouette of a sun-glinting Cessna 195. We all had departed from Schellville Airport and Submarine Base, a unique and sometimes submerged airport, one of San Francisco's gathering places for vintage aircraft.

All this is done in the gentlemanly manner of northern California, all with restraint and good taste, not to be confused with the surfing life of southern California. One of the first things I learned as a guest in the Bay area was not to be shocked at the spoken wish that the next major earthquake would snap the state in two, south of Monterey, and the southern half and all its manners and mores be allowed to sink into the sea.

We were plodding across wine country, the valleys of claret and Chablis. They have gourmet food for transient aviators out here, the Flying Lady in the valley of San Jose, and the famed Nut Tree airport just ahead. I was lost, but my host, pounding on the cowling, pointed to the Nut Tree runways just forward of the lower-wing leading edge. Too high, too close. He signaled with his hands that a couple of falling loops should position us just right for landing. They did, and still full of the freshening experience of chain looping the Stearman, I had lost all my finesse, and used up both sides of the runway coming in.

Refreshed with a snack and delicious coffee, we once more climbed into the worn golden hills of California. Once, cresting a ridge, I caught a flash of something white in the corner of my goggles. It was an eagle, a magnificent soaring eagle, riding the wave at the top of the ridge.

And in seeing the eagle, I saw something else. There below the eagle was the moving shadow of a sailplane, it's thin wings riding the same lift curve as the eagle above him. How beautiful.

Then, a shadow fleetingly crossed my open cockpit, and I looked up, startled to see that we were flying in the aluminum overcast of a giant C-5A descending toward Castle AFB. The C-5's pinions were spread and fluttering, like the eagle's. There we all were, jet transport, Stearman, eagle, and sailplane—all silently using the same airspace for different reasons. Layered, watching each other. California flying.

Another thud on the cowling broke my reverie. My host was pointing to some scratches in the desert below. We were at Merced. Not a city, not an airport, not like anything else in the world.

I never did make a really slick landing or takeoff to show Lee how good I was with his Stearman. That happens to me all the time.

There was a good breeze among the rows of antiques parked at Merced, and they said the temperature was 105° F, but it didn't seem hot. Back home on the Gulf Coast, 105° F would broil you in your own juices. Here I didn't even sweat. That's a bad sign. That means your liquid is evaporating before it ever gets to be sweat. I thought the climate was wonderful, and I began to feel more and more wonderful. Somebody said, "You'd better get Bax a hat. Look at his eyes."

My eyes were looking at airplanes I never thought I would see . . . and at California women. Ain't a flat one out there, near as I could tell. And I was looking in vain for the commercial exhibit area, for the boys in the Wichita blazers. None of that either. Cal Chaney of the Merced Pilots Association said, "Don't try to make it something it's not. This is just a bunch of us getting together every year to look at each other's airplanes. This is the first year we've gotten far enough ahead to serve beer and chicken."

Innocent. They are all innocent. I strolled among the strollers, looking at airplanes too.

Good Lord, I never thought I would ever see a real live Gee Bee. "Me neither," said Bill Turner, "so I built one."

"Sure looks like it would make a big blunt hole in the ground," said I.

"It tries," said Turner honestly.

There was always a big crowd around the rare, mirror-polished B-23 bomber that used to belong to Howard Hughes. Whether they were admiring it or just copping shade under its big broad wings, I don't know.

And there was an honest-to-God Pitcairn Mailwing. I thought they were all in museums. This one flew in.

And there were the bones of the Gonzales #1, built in 1910 by Willi and Arthur, the Gonzales twins from Los Angeles. Nephew Bob Gonzales, who found the skeleton, dry and preserved, says he will make it live again.

Mere Stinsons, Wacos, and Staggerwings were commonplace. The little Aeroncas, Taylorcraft, and shiny Luscombes were separate, considered only "neo classics" in California.

I kept going back to the Boeing P-12 pursuit ship. Lew Wallick could see what the little biplane was doing to me. "Go on, sit in it. Here, put on my helmet and goggles." I asked him if he would take a picture with my camera while I imitated a pose I had seen of Lt. Claire Chennault when he flew this fighter. With a touch reserved for . . . well, not airplanes, I placed my hands and feet on the stick and rudder, looked out past the N struts, and had my own fantasy. This is really what I should have done in my life, fly the P-12.

Lew may be the happiest pilot in the world. Chief test pilot for Boeing, he keeps this oldest flying Boeing as a pet, and at night they let him tuck it into the hangar at Seattle to sleep between the legs of a 747 at the factory.

Oh, that stubby, long-legged, reared-back little fighter and its competitor, the Curtis Hawk P-6E, were probably two of the most beautiful biplanes ever made. Once, when I was doing a story on Executive Jet, a Lear charter outfit run by Paul Tibbets out of Columbus, Ohio, I saw a picture of the Curtis Hawk on his wall. What was it like to fly?

"It flies just like it looks," replied the former pilot of the *Enola Gay*.

I asked Lew Wallick the same thing. Got the same answer. Then he added, "Of course, it's hard to land. It was supposed to have a tail skid back there, dragging a straight line in the turf, instead of a castering tailwheel."

Wallick had to leave early. I shadowed him while he suited up, stood aside, and listened as his big radial coughed to life. Then I trotted alongside the little fighter all the way out to the runway just to be close when he took off. Three plane lengths, and in a three-point attitude the Boeing P-12 was off and gone.

"Why you going home so early, Lew?" I had asked.

"You see those exhaust stacks? Poking straight back like bayonets behind each cylinder? Well, flying behind that thing at night is just like flying behind Halley's comet."

Drifting back down the flight line I stopped to admire a Breezy. "You ever fly in a Breezy?" the owner of this one asked me. He was a big old senior captain with United, mustache flowing, bugs hitting his teeth. He strapped me in and flew that contraption with the same care as if he had 300 souls sitting behind him. He let me handle it some. An honest airplane and a view of all the earth below—just past your shoes.

Later I asked him why such a lightly loaded birdcage needed so much power on approach?

"If you wanted to invent a machine to produce maximum drag you would use an open framework of round tubing. Like the Breezy."

Pilots at Merced are like that, likely to take you for a ride if you look like you need one.

After the sudden and flaming desert sunset, everyone drifted in to gather at the long tables for food, drink, and music. The room was gentle with laughter. "Orville, how did all this get started?" I asked.

"Well, a little bunch of us kept some Wacos here. We liked to socialize and go to fly-ins, and decided to have one of our own. Me, Pop Fleming, his son Bob, Olin Starnes, George Dray, Tom McLendon, and some of the other locals got it together in 1958. We put some hand-lettered posters around in airports, and the only response we got was one lady who flew her Moncoupe in . . . We weren't sure there was such a thing."

Gail McCullough, who now runs the public relations for the Merced show, said they had registered over 16,000 people this year (1979).

Next day, bumming a ride back to the left coast with Greg Walker in his 172, we flew over the dreamy golden hills, and I asked him why Californians go so deeply into their hobbies without a trace of commercialism.

Greg said he didn't have a why for it, but wait until I see Watsonville and Livermore, Porterville and Columbia, and all the rest of it.

I didn't have any idea all that was going on out there beyond the Rockies, did you?

16

The New Boy

THIS IS ONE OF THE FEW DIRTY TRICKS ever pulled by Eagle Scout Collins and the Editors of *FLYING*. Convinced that the lack of new starts in student flying had its beginnings in the apathy and closed-club attitude of the average FBO towards an outsider, not to mention the lack of advertising and promotion of their businesses locally, the staff sent me out to prove the point.

Not good journalism, this concept of already having the story in mind and setting out just to prove what we believe is true, but it worked here. I did the field research in the spring of 1980, and we carried the story in July under the title of "First Encounter With Flight, or Even With Money In Hand, Getting Into Aviation Can Be A Challenging Task".

I came into the airport from a door on the street side, wearing suit, tie, briefcase, my bifocals down on my nose to accent my age. My story was that, like any businessman who'd heard that flying his own plane might help business, I wanted to find out for myself. I would always stand ill at ease, peering around in the airport office uncertainly until somebody spoke to me. Sometimes they never did, and I would take the initiative with the foolish sounding "Er, ah-h, who do I see around here to learn how to fly?"

I did this at several locations across the country, at small-town grass airports, at middling towns with good paved airports, and at big city internationals with jets roaring in my ears. My aim was to compare the orderly businessman's world to the world of general aviation. The comparisons were not good.

I had rehearsed my cover story. My home was in a neat suburban neighborhood; everything in it was comfortable and served me well. My competent office staff was minding things while I was out on this trip. Some of my associates at the country club had acquired small airplanes. I knew they were expensive, but my acquaintances seem enthused about flying; they spoke well of their airplanes. I had read some aviation magazines before going out shopping for flying lessons and perhaps buying an airplane of my own, but all my previous experience had been only as a passenger on airlines. One reason I was there was the increasingly unworkable airline scheduling.

At each FBO I visited I trolled this delicious bait before the ones who would even talk to me, and not all of them would talk to me at all. Wherever we got as far as exchanging names, I introduced myself as Hank Snee and explained that I had no business card yet as we were just relocating there. I had offices in nearby cities and needed to travel among them where airlines no longer served. Wasn't that dishing it up on a platter?

I look sort of Hankish, and was wearing my brown snee coat. That part of the story always worked good wherever I got far enough to get anyone to listen. I was worried about the demo ride, fearful I might make some little giveaway movements, like looking before I turned or absentmindedly centering the ball while I was overcontrolling the airplane in a manner I envisioned a first timer might do. This concern lent credence to my general attitude of worry.

My bait speech was so good that one or two of the larger, better-organized FBOs went for the quick money first. Instead of responding to my request to learn to fly, they made a hard pitch for their own charter services. And nobody caught onto my tyro flying act, which gives me second thoughts on how easy it is for me to fly that badly.

Most important of all, nobody ever tried to sell me anything first. And not once did any of them, large, middle, or small, offer me a demo ride, not even after I had cooly said that the average $2,000 price for my license seemed reasonable. ("But I sure would like to see for myself what one of these little airplanes is like.") No, I always ended having to ask for a demo ride and offering to pay cash for it then and there.

With few exceptions the general aviation industry talks to itself. I sometimes got a very good pitch, but it was only after I had taken the initiative to break the ice. My unvarying impression of trying to buy something from an FBO was much the same as walking uninvited into a private club.

I can think of no other industry so isolated unto itself or so unaware of the basic "meet 'em at the door" art of retailing as our own beloved general aviation. I had never noticed this before because I had been on the inside myself for so long.

There are a few really good presentations worked out by the manufacturers. At the field level, the pilot training programs were mostly well-presented and attractive—once I got the FBO's attention. But there were far too many cases in which the FBO relied entirely upon the manufacturer's kit, and sent the prospect off to study at home.

Let me tell you some of the horror stories. At a clean, good-looking FBO they wanted to sell me the $65 "kit" and send me home with it.

"But I sure would like to find out what flying is like. Is it all that good, or will I be scared or get airsick? I don't know . . ."

"Well, we are all too busy just now. Here is my card."

Once, I flashed Hank's wallet, loaded with greenbacks, and asked if there was anybody there on the field I could buy a ride from?

The youthful CFI I was talking to made a trip back down the hallway, and I heard the chief's gruff voice say, "Tell him it'll be at least a half hour before I can fool with him."

More leaving motions from me cut the time down to a quarter hour. I cooled my heels looking at vintage aviation calendar art; nobody invited me to sit down. The chief finally sent the kid back to sign up Mr. Snee for a half hour in a Tomahawk for 25 bucks. The kid looked peeved. On the way out to the flight line he stirred himself to tell me that the Tomahawk was a little crowded and dual instruction in it cost $47.60 an hour. For only $54 an hour I could learn to fly in the bigger, nicer Warrior.

Ol' Hank said he didn't know a Tomahawk from a teepee.

That was true too. I had never flown in one. It was easy for Mr. Snee to say, seeing this machine for the first time before trusting his life in it, that it sure was pert and attractive looking.

"Sure is a nice-looking airplane."

No reply.

The kid did his walk-around out loud for my sake, then tried to hide his anger and confusion when he couldn't get the left door shut from outside. He gave up, got in the right seat, reached across me, and latched the left door.

My first ride in a Tomahawk. Boy they sure don't want you to miss that fuel selector switch do they? It hung off the center panel like a cut-open watermelon. The kid brushed off my questions about that. He was doing the runup and was concentrating on leaning the mixture out as far as he could to "burn off" the right mag which had a 250 rpm drop and was barking like a fox.

Ragged engine and all, the little Piper flew eagerly, pleasantly. He gave it to me once we were in level flight.

"You don't have to grip it like that, sir."

I was showing him plenty of white knuckle, the yoke in both fists.

"Look, it flies by itself."

He put the Piper into a gentle left turn, took his hands off, and it stayed there, perfectly balanced as it kept the same turn rate. I didn't know they'd do that.

It was hard for us to talk because he had left the storm window open. After about fifteen minutes he reached across me and smartly shut it. "Guess we've got enough fresh air now that we are up at altitude," he lied.

But he never mentioned the rudders. We made the whole flight with feet flat on the floor. So did all the others, but the good ones pointed out that rudder pedals were there and would be important when I later learned about coordinated flight. Both the Piper and the 152 flew just as well with dead, ignorant feet.

At the big-city FBOs where they used Cessnas, they told me to fly with my left hand only and to just hold the yoke with my fingertips. There were two good instructors. One was a furloughed airline pilot who confided to me that he disliked teaching but loved flying—and also his wife and kids—and for now this was all he could do. The other was a smooth 1,100-hour pilot who told me he was just passing through on his way to a corporate flying job and this was the only way he could build time.

But regardless of their dim outlook towards flight instruction, neither of these men would let a mistake of mine go uncorrected and unexplained. They did it from habit, I guess. Ol' Hank would have gone back to either of these men to learn to fly. Little as he knew, he knew they cared.

One instructor gave Hank a long demo ride, then let him fly all the way back, taking over just inside the pattern of the busy reliever airport. On downwind, tower cleared a Baron for a left turn right in front of us. I could read Goodyear on its tires and Bendix on its brakes.

"They sure fly close here." I wanted to see if that Baron had scared him bad as it did me.

"Tower shouldn't have done that," said the demo pilot through his teeth. And later, searching for honesty and trying not to scare Hank off he added, "That was close, but not *that* close."

Sometimes, if the demo ride had gone well, I would ask if I could try to land it? The near-miss pilot brushed my hands off the controls and said, "There are many basic things about flying I'll need to teach you before we do that." Nearly all the CFIs were totally honest once we got aloft.

Hank Snee would have been attracted to the big city FBOs' clean and well-maintained aircraft once he got into one, and the honest presentation of the $1,400, six-month, pay-as-you-go package at the Piper stores—$1,500 at Cessna (1980 prices). All the big ones took pains to caution me that the package price was unrealistically based on a 35-hour minimum flying time and that it often took longer than that. Some of the middle-size FBOs were just as candid, explaining that individual skills, weather delays, and rising fuel costs could easily run it on up to 45 hours before I was "safe." All of them used the word "safe" often.

The painful part of this story deals with the segment of aviation I have always instinctively aligned myself with in the past—the little guy, the romantic grass roots flyers, out there in the meadows by small towns.

Again, I was unaware of what snobs we are. I have always been a part of this rusty-hangar and wildflower flying, shying away from, and feeling uncomfortable with, the pilot lounges and slick offices of the big-city FBOs. But now I was to see it as Hank Snee would see it, and Hank was not born in a worn leather jacket; he was not part of the brotherhood.

Hank wandered onto one placid airport. There being no office as such, he walked into the hangar. Ol' Junior was sitting in the hangar mouth, enjoying the breeze, watching the traffic come and go, two black cocker pups playing at his feet. Junior refused to shift his distant gaze or acknowledge that I was standing there in my suit, holding my briefcase.

"Nice pups you have there."

"All they do is eat."

Junior hadn't done so bad himself at eating, judging from the bulge of his bib overalls. The bowed legs of his aluminum chair strained under Junior's grease-stained bulk.

Silence descended between us again. Junior had not found me worthy of a glance.

"Where could I learn to fly an airplane?"

"Right here."

More silence, and the distant music of a faded green Cessna doing touch-and-goes.

"Cheapest place on the field," offered Junior. "That's what they all tell us. We're the cheapest."

Hank looked all around at the gentle decay of the airplanes and hangar. The faded signs, flaking paint, rusting tools. The man was right, no doubt.

"Well, could I talk to somebody about getting started?" Hank made little starting motions.

"Gotta wait till Doc comes back. I don't instruct any more."

Doc, it seems, would be back in about two hours. Or I could wait and fly with Mitzie. Junior's voice didn't sound the same about Mitzie. She would be back in about an hour. Hank Snee was tempted to wait and get a look at Mitzie.

Instead, Hank asked if he could look at a training plane? "Sure. There's one out yonder."

Out yonder, in the direction of Junior's nodded head, lay a Cessna Aerobat in the high grass. Its gallant checkerboard paint long since gone to the sun and wind. I waded out there, past some Cherokees and a weary-looking 172, all equally sun-crazed and silent. Any dual time in that Aerobat was, first of all, going to involve somebody finding the missing right seat. It had bald tires, and about a gallon of vinegar might have washed the thick mold off the plastic inside. You could have bedded tomato plants on the loamy floor.

Hank Snee shuddered and backed away. Junior never got up as Mr. Snee, about two thousand bucks on the hoof, slowly walked away.

At a lively but small FBO at a middle-sized town's airport, there was again no way to find the "office." Snee walked in through a hangar that would have delighted Bax: a turkey-tailed Bellanca, a wood-winged Mooney, and a noseless Beech 18, congealing in its own tar. A silver Cub stood tipped on its nose against one wall to allow room for the bass boat and trailer.

A desolate push-pull Cessna carried its own messages written in the thick coat of fuzzy grey dust on the cowling.

"Wash me."

"Fly me."

"You've waited too long."

Beyond a Ryan PT-22 against the other wall were the absolutely fascinating remains of a Decathlon that had spun in under power. Only the tail feathers, stacked on top of the crushed and blood-stained cabin, identified the mess. The engine and instruments were flying somewhere else by now.

Bax stood in morbid fascination: a pilot looking at a really bad one. And Bax wondered how Mr. Snee felt seeing the insides of a crashed airplane.

It was the kind of a hangar and airport that all my best flying tales have come out of. But I let Mr. Snee find the office, go in, and just stand there until spoken to.

The old pilots, my natural friends in real life, just looked up, discounted me as a "townie," and went on with their family talk. A young instructor and some students came and went. Snee put his briefcase purposefully up onto the countertop with a light thud, and waited. He studied a big hand-lettered sign advertising flying-lesson costs and rental rates for the different airplanes. A receptionist had gotten her fat parts snuggled way back into her office chair, the phone tucked under one ear against her shoulder. She was humming along in a murmured conversation with someone far, far away. She ignored Snee.

Finally, a fine-looking woman with iron grey in her dark hair came out of a back office, looked directly at Snee, smiled, walked right up to the now stiffening Snee, and asked if she could be of any help.

It is to the everlasting credit of this handsome woman that she gave Snee the entire presentation of the Cessna Pilot Center training program more briefly, more comprehensively, with more warmth, and in less time than it had ever been told to him before. Some of the big-city FBOs made an overly elaborate presentation of the Wichita brochure, wasting time with a canned sales pitch on a man who must be already sold or why would he be standing there in their store?

Other desk-front monologues took an hour, dragging on into the confusing comparisons of audio-visual cassette study at home, supervision at the airport, and enrollment in the on-going FBO ground school classes. Once they got started, some were pretty aggressive. One tried to sell the as-yet-unflown Mr. Snee on buying a 182 when the customer had no idea of the difference between a 182 or a 172 and very little idea of what the salesman was talking about. The salesman smelled money.

In contrast to these scattershot proposals, the good lady at this middlin' airport laid only one piece of paper on the desk, jotted down the current prices so Snee could take it with him and think it over, and—like all the others—warned him that the FAA-required hours were only a minimum. Learning to fly safely might take a little longer.

But again, it was not until Mr. Snee said he sure would like to take a demo ride in one of the training planes, maybe even today, that the lady flunked out. She pointed to one of their 152s sitting in the grass and told him to help himself. He could look at it all he wanted to.

I went out and spent a lot of time looking at the little 152, like I had never seen one before. I could feel the eyes watching me from inside that comfortable building, but nobody ever came out. When Mr. Snee at last hefted his briefcase and $2,000 walked slowly away, nobody made a move to stop him.

This one was a non-wealthy operation but with good airplanes, but just letting a customer walk off without trying to fly him is something not even a little Ma & Pa grocery would do.

I knew from my own experiences that some of these old nesters at the little airports could teach a new student more common-sense flying than all those coat-and-tie big-city operators combined. But Snee would have long been "that new guy," long after he had been introduced around and his cash started to flow.

I had a problem of my own in flying my Mooney to all these places and not having my cover blown by having somebody notice I had driven up in my own airplane. Most airports have more than one FBO, so I would park at one, ask them to top it off, and walk over to the other. Nobody noticed. I got bolder, and once, at an airport with only one FBO, I left the Mooney down at the far end of the ramp, walked around to the street side of the hangar, and came in as Hank Snee the tyro. It didn't matter. Nobody out front ever asked me if I wanted to buy gas, and nobody in the back asked if I wanted flying lessons. I was simply not able to disturb the tranquility of it all.

Once I flew with a college youth on an asked-for demo ride, and he was open, friendly, and explained to Mr. Snee where flight instructors come from.

"I only have a few hundred hours flying time myself, but it's not all that hard to get to be an instructor. Teaching is mostly pretty boring, but it's the only way most of us can build up enough flying time to go on to better jobs. I just love flying . . . ."

That left old Snee with a lot to think about—the new experience of trusting his life to a small airplane. What the kid said was absolutely a fact of life in general aviation. The only thing he did wrong was blabbing it out to his scared prospect.

Once Snee could get aloft, the quality of airplanes and instructors was, for the most part, satisfactory. In fact he took a liking to some of the guys that toured him around, and ended up feeling rotten about his own sham. Suppose the guy made a really good demo ride, and yet the customer never showed up again? He needed to know why. What happened. That it was not his fault. So I deliberately blew my own cover a time or two with some of these guys. Here is the funniest one.

It was with the furloughed airline pilot, and we were in a 152 coasting in on final at a middle-sized controlled airport. "May I try that?" asked Hank Snee, who had done pretty good flying level and making gentle turns for the first time in his fictitious life.

"Okay. You want to hold the speed at 70, right here," tapping the airspeed indicator. "And try to aim it right down the centerline of the runway. It will slowly descend by itself. Keep the wings level."

"Which one is the runway?" asked Mr. Snee, sliding a hand onto the throttle shank, noting the instructor poised and ready to snatch it all away.

Well, flight instruction *is* boring, so why not put a little fun into this? I set up a delicate forward slip to exactly compensate for a slight crosswind and gently milked the throttle back to hold 70 for once in my life. Just before the sweetest flare and kisser landing you ever saw, he leaned forward, peered into Hank's bifocals and now-grinning face, and asked, "What the hell's going on here?"

On the rollout I leaned over and shook his hand. "I'm Bax, *FLYING* Magazine."

"Sonuvagun!" he cried, both of us bursting into laughter.

Taxiing in I told him what it was all about and that he had made an A+. He was still laughing and shaking his head.

After this story was published in the July 1980 *FLYING*, we got the expected letters of outrage from FBOs and flight schools who *do* greet people at the door, *do* make strangers feel welcome, and *do* advertise locally in their own markets. They hotly resented our story. Without exception each of these was a most successful and profitable aviation enterprise.

17

Justifiable Scud Running

> **scud runner / skud-runner** *n*, E *scud* dirt, refuse; a blend of E scum and mud + ME *rinner, renner,* fr. *rennen* to run: a pilot who inserts his head into a narrowing slot between ground and uncertain ceiling with hopes of arriving at destination airport before the blade drops on his neck— aeronautical slang derisive but not necessarily contemptible

I HAVE NEVER FELT the same good-humored laissez-faire about scud running since reading a story called "The Short Happy Life Of A Scud Runner" in another magazine in which they printed the last words from the cockpit voice recorder. The airline Convair was playing in and out of clouds in the Ouachita Mountains of Arkansas a few years ago. The last words were the co-pilot reading from the IFR en route chart, "Minimum en route altitude along in here is forty four hun . . . . ." (sound of impact).

There was a time—in my younger, freer days, before radios were common in the little planes we flew—when I considered any slice of sky that was still open to be legally mine. I knew that nothing tall grew on the interstate highways, and there was nothing as comforting as the rain snare-drumming on my windshield while I steered the little Aeronca homeward behind a steady stream of cheery red taillights.

Home, across the river. Flying shoulder-high to our modest downtown skyscrapers that were huddled yonder in the mists. Cutting across the garish neon lighting of the shopping center. Curving onto the path of the blacktop that would lead me to the edge of town and out to the grass airport. The airport was easy to find in the gathering gloom; it was the only big square of town without any lights on it. If you nosed-in low enough, there was a glimpse of the hand-painted white numbers at the end of our only paved runway. Ireland couldn't have looked any better to Lindbergh than this sight did to me.

A careful jockeying of the stick as gusts and rain tried to have one last fling at me, then tires rolling in a spray of water, and slowing down—home. An exultant dash from under the wing after tie-down, then on the phone to the wife. "Once again I have cheated fate," I would traditionally cry.

"You idiot," came the traditional reply, "bring home milk and eggs." Poor use for a hero, but I was thankful to be doing mundane stuff among earthlings again. Looking up at the still lowering sky, I felt both grand and guilty for getting away with it again. Now just another taillight, backtracking the same blacktop that had brought me in.

The years went by. We bought a Mooney and I went to an expensive school and learned how to fly it without looking out the window. Soon we found that the 1968 instruments in the old Mooney were not adequate for serious IFR flying; they were raked out and replaced with new instruments, six pounds of them at about $2,000 a pound, which got close to making my instrument panel more valuable than the entire lovely little flying machine. And I thought that was absurd.

All these jewel-like instruments were designed solely for the purpose of flying in clouds, for flying when you can't see out the window—which to me is like renting a cabin with a gorgeous mountain view and staying inside all day with the blinds drawn and watching TV.

But I really felt I should use all this hard-earned cloud flying knowledge and equipment, and for a time there I went out of my way to do so. I got to be safe at it, but never really good at it. I just didn't like it.

I am not one of these aviators who enjoys such precise flying as to draw the bright blade of his instruments against the wooly roiling cloud. And, anyway, sometimes it scares me. I have had my ashes emptied and my aluminum stretched by being vectored right into the middle of a seething little embedded honeybun of a storm cell that I never saw—all this while flying the "system" under the guidance and auspices of an agency of the United States Government whose published reason for existence is to promote and enforce safety in flying.

I am not speaking of those bright, beautiful, on-top instrument trips that I could not have otherwise made and which consisted only of a punching upwards through a layer of overcast at the beginning and drifting downward through such a layer at the end. I am speaking of those long grey journeys, my mind fixed on the problem of keeping the shiny side up, arriving brain-and-butt-weary to come down the localizer, all furrowed of brow, and breaking out of a thousand-foot ceiling to find you can see forever once you have gotten beneath the lid. At such times I always think of the good old days of just following the freeways. I wonder why I feel so guilty about the idea of scud running now? Well, I used to feel guilty about it *then*. I always used to feel more like a creeping barefoot house burglar than an aviator.

And so it came to pass that I was weathered in; it was the springtime of the year 1980. I took my anti-get-home-itis pill and resolved that all my worldly affairs could wait another day. I fell in comfortably with an old friend, a Baptist preacher, and decided to wait and enjoy while one springtime front chased another across Texas.

But it nibbles at us, you know. It nibbles at us. Thinking of my Mooney tied down out there, burdened down with all those expensive weather instruments. The airplane itself does not know nor care what the weather looks like unless you enter something severe that tries to take its wings off. The airplane out there was willing to have me home in an hour and a half. At noon I checked weather again. I was restless. The preacher went to his office. He said he would speak to his Boss about prying open the sky a little if that would make me feel any better.

I was at Austin, in central Texas. West Texas was still high overcast, but with the hooves of a herd of SIGMETs galloping this-a-way just hours from now. The storms were crossing the Rio Grande. Weather was not bad at Austin, but to the east, toward home plate at Beaumont, the sky slanted downward into rain, fog, and embedded cells. The usual Gulf Coast forecast.

But the first half of the trip was calling for 1,700 feet, light rain, and scattered thunderstorms. An old Dick Collins saying arose in my breast, "No harm in going out to look at it. Plan all the places where you could quit and land. No matter where you give up, you will be that much closer to home . . ."

Weatherwise Collins knows, as so many old pilots do, that freaky weather opens, closes, moves in, and moves out much too fast for any agency to tell us about.

I launched but did not file IFR. Why not? Those imbedded cells, those loaded dice up there where a real instrument pilot would be forced to play, that's why. I decided to just scud-run it, to play what an imaginative pilot could call VFR.

My handicap in this game included low flying and limited radio and VOR reception. And also knowing that my back door to Austin would be closing right behind me.

My hole cards, and I would never play this game without aces in the hole, were a flat featureless terrain, knowing the way, and knowing the string of airports along the way where I could quit if I had to.

Through rain and beneath the scud, admiring the wet farms below while a fortune in IFR needles lay unused on the panel, I called ahead at each waypoint that had a radio. The last one reported my current home-plate terminal forecast as a brief 1,700 feet and rising, light rain, and three miles.

I made a mental note to call the preacher and thank him. I wanted to tell him the Mooney is only seven feet tall at the rudder, and 1,700 was prying the sky open plenty far enough.

It had been many years since I had come home in low rain along the blacktop road. Tuned to Approach over at the big airport, I could hear some guys fighting their way down the ILS. I felt as if I was stealing. And home plate shut down tight right after I got home with the milk and eggs.

Conclusion: moderation in all things, sir. Habitual scud running, especially in the high country, will get your picture in the local paper.

But the weathering sky appears in an infinite variety, so think of legal scud running as one of the interesting ways you can respond.

18

Skydiving Cats (Part II)

I KNOW BETTER THAN TO DO THIS, but pilots keep coming up to me, on the flight line or at some banquet where they have hired me as speaker, wanting to know when we are going to finish, as promised, the report on skydiving cats. I can tell by the silly crafty grins that this is what they are going to ask me about even before they do it. There is no place to run, no place to hide.

We ran the original story back in June 1980, and the mail surfaced more cuckoos than you would ever believe. That was when I decided to let the subject of skydiving cats, pardon the expression, drop.

Then, at our last editorial meeting, my own editors, Collins and Brechner, said to me, "When are you going to finish the story on the skydiving cats?" This from serious responsible men who direct the future of *FLYING*.

For those of you who might have been out of touch, let me recap the story up to now.

In a letter of January 1980, Bob Hamp, then of Roscommon, Michigan, told me of some old nesters who fly from the isle of Bois Blanc, high in the cold north country in the Straits of Mackinac. The story centered around Allen Hoffman, a then-80-year-old crusty ex-barnstormer, airport manager, and beloved teller of tales from 'round the red hot stove on winter nights, nights when only the whiskey doesn't freeze.

"These fellows fly an old Champ; they carry the cat up to something like 3,000 feet and launch him through a piece of six-inch stovepipe fixed to the side of the fuselage for that purpose. I'm told the cat is waiting to jump again when they land and taxi back to the grass ramp."

You will note the factual vagueness of some parts of this story. While we have the altitude and location of the operation, and the diameter of the stovepipe, the operators are only "those fellows." Many other questions are left unanswered in this cat tail, er, tale. Does the cat wear a chute? A long-haired or short-haired cat? Tomcat or minnie cat? In my inconclusive report of two years ago I promised to get the facts. To find the terminal velocity of a cat, possible effects of density altitude, any FAA rulings, and an opinion from the Society for the Prevention of Cruelty to Animals.

All I ever got was all this silly mail. And a lot of uproarious laughter from my local FAA GADO office, which kept transferring my call from office to office. I finally got Carl Edminson, Accident Prevention Counselor, and when I told him it was not an accident, they were doing it on purpose, he said to call him back Monday. Monday of next year.

The FAA wouldn't stop laughing long enough for me to find out if you need an STC to stovepipe the side of a Champ (would the airplane then be "experimental"?). One fellow said he thought the cat was experimental.

I got an early, firm answer from the SPCA. They said, "Any cat who persists in hanging around small airports and in the company of pilots of light airplanes is beyond concern to us."

My only positive response was when I promised that I would try to find some photo-jumper with enough hair on his bod to bail out with the cat and free-fall long enough to get us some air-to-air studies on film. Several enthusiastic young men replied with photos of themselves doing unimaginable things in free-fall. One said he would jump out of anything that flies, eat anything that moves. All were from California.

But the rest of the mail! They wanted answers. One, addressed to "Flaying Magazine" from a gentleman in Missouri, said that if we did not print the second half of the skycats story that we were dead and gone as publishers and it would be three generations before any of my descendants could come out of hiding in the swamps.

A pilot from Long Island told of taking off in his '52 Piper Pacer and hearing a seatbelt dangling outside, about to beat a hole in his fabric. With full flaps he slowed to about 55 mph, and in the opening and closing of his door, his eight-year-old Siamese, "Cognac," sprang out the door at 1,100 feet. Right over the field. Hurriedly he landed. "Cognac" was sitting by the runway, washing his face. Unhurt.

The story got around the airport and his mechanic asked if he was planning to drop a cow from a DC-3 next. A hundred-dollar bet was quickly made that this pilot could not drop a cat unharmed from 1,200 feet. By drop time on Sunday this pilot had $1,400 out in bets and a good crowd. He dropped a grey-and-white slick-haired stray, not wanting to risk "Cognac" again, even for big money.

The stray sailed over the heads of the onlookers, missed them by about 200 feet, got up, and headed for the woods with a speed and agility that indicated outrage but no injuries. The writer of this letter said he was considering catdiving as income to pay for an upcoming engine overhaul when another pilot had two feline fatalities the following Sunday. "We all felt pretty embarrassed for participating in such a stunt."

Another reader, from Texas, sent a clipping from an April 1934 issue of *Popular Aviation*, ancestor of *FLYING*, proving that nothing is ever new. The cat launched from a Curtiss Robin, complete with harness and three-foot-diameter parachute canopy. The chute carried away from the harness on opening, and the cat survived a free fall of 1,000 feet. ("Although stunned," read the official report.)

My mail included photos of a 150-lb Newfoundland dog, standing in the doorway of a parked twin Cessna, wearing a harness and chute. This Florida reader said "Tuffy," the dog, was a regular member of their jump team, although some pilots refused to carry the big dog aloft because he drooled on the carpet. I thought all skydivers did that.

A Connecticut writer told of the parachuting bulldog mascot of the 82nd Airborne at Fort Bragg, North Carolina, in 1950, and how the rival airborne division over at Fort Campbell had a harness and chute tailored for their mascot, a monkey. At 3,000 feet, man and monkey went shouting out of the jump plane. "What more beautiful camaraderie of man and beast in mutual endeavor?"

The only problem was that the monkey liked his chute so much he climbed the suspension lines to ride the canopy on top and went in as a streamer. By order of the Division Commander, no more four-legged animals in aircraft.

Nearly all the writers asked that their names be withheld. Some were simple stories of a pilot who carried his pet cat and dog as companions until one day his cabin door popped open in flight, the cat jumped out, and the fool dog chased the cat.

I got several of the oft-told tales of the man who claimed he never learned to fly by instruments but got along okay in IFR weather by carrying a cat and a dog. In cloud he lets the dog out of his cage, the cat's tail bristles up, and the guy flies by reference to the cat tail instead of a turn-and-bank indicator.

A fellow cat lover, a PhD from California, wrote of a cat with a chute whose own self-righting gyroscopic-like capability proved fatal. The reluctant cat was launched in a tumble and rolled himself into a fouled chute. The writer said the final experiment was with a large, white, leghorn rooster out of a 172 at 2,500 ft. The skydiving doc went along as observer.

"He went nose down, full delta until near the ground, then deployed brakes and flaps which came unglued. He creamed in. There were still feathers in the air when I landed . . ."

The Doc said he had never witnessed a successful animal jump. From the stories I got, cats seem to run a better average of living with no chutes. Or maybe that just makes better lies.

My cat is staring at me balefully as I write this. Cats know what's in your mind. I decided to make a wrap-up and withdrawal from this whole thing when ABC-TV's "That's Incredible" called me at home and wanted to know if I had done the story and would I give them an interview and some other names and addresses?

After watching the "60 Minutes" treatment of general aviation as an example of what big-time network TV can do, I decided that I do not want to be remembered as the guy who started the "cat splatting" craze in America.

And I would warn any of you who are still captured by the idea of skydiving cats that God may turn out to be a giant Siamese tom, and the first thing He's going to ask you is, "Did you ever throw any cats out of airplanes?"

19

Braniff's Final Run

FOR THE OCTOBER '82 ISSUE of *FLYING* I wrote about what it was like aboard the last flight of Braniff International. The person who told me most of this story is Carolyn Brown, called Charlie Brown, a beautiful woman and a senior hostess for Braniff. She is my wife's best friend, a friendship that began back in the Convair and DC-7 days when Diane flew the line and she and Charlie Brown had adjoining apartments. The Braniff crews have often been compared to a big family: Diane and Charlie are like sisters, and Charlie's husband Bill Rees is flying again now for Continental. Braniff has struggled back to life again, too.

But on the night they told this story they were all gathered at our kitchen table like a stricken family, all fresh out of flying jobs. There were forty cups of coffee and a million tears. Charlie Brown had been one of the crew working that last flight home.

****

A little known story is the legend of Fat Albert and Captain Charlie Lamb the day the company went broke. Charlie Lamb was one of the best liked senior captains with Braniff; he was due to retire in just a few months. He was passing over Los Angeles outbound on the company's prize run, Dallas to Honolulu nonstop, when he got the call to return and land at LAX. Braniff President Howard Putnam had just concluded that the company was broke and was trying to get all his chickens back to roost at Dallas rather than have airplanes scattered all over the world to be picked off, one by one, by creditors. Putnam had them all home except his one-and-only 747, lovingly known by company and customers alike as "Fat Albert." Braniff had other 747s on lease at the end.

Time, in their story of the demise of Braniff, was so sure that Fat Albert had returned as ordered to Los Angeles that they wrote their story like that. There was no way *Time*'s reporters could have known what was actually happening on the flight deck of that 747 nosing out over the broad blue Pacific.

The co-pilot said to Capt. Lamb, "They want us to come back."

"Tell them I can't hear them."

"They've ordered us to land at DFW, Captain."

"Tell them I can't hear them."

A long silence on the flight deck, then Charlie Lamb turned to his crew and said, "Listen, those folks back in the cabin bought a ticket to Honolulu, most of them are pretty excited about a vacation like that. We've got plenty of fuel, beautiful weather, and a good running airplane." Captain Lamb paused, "I'm taking them to Honolulu. What can they do, fire me?"

The dreaded news had already spread among the airlines. One 727 crew got the news from Miami Center, inbound from their South American run. "You boys know you are working free right now, don't you?" was Center's gentle way of putting it.

Putnam got all his airplanes back but his Honolulu-bound Fat Albert. Other airlines began to ferry stranded Braniff layover crews back home from all over the world. No tickets, no cash. "Just wear your uniform. We are not counting heads this trip."

The yet unresolved story is that they all did this except American. I've had Braniff crews tell me American Airlines cold-shouldered them. I've had American Airlines officials tell me it just ain't so. There is no question about American doing all they could to put rival Braniff under in the last days. The animosity surfaced now and then. My Diane had to make a trip to Dallas to see her dying father. Her best suitcase still had "Braniff Crew" markings on it. On a commuter trip from Beaumont to Dallas her bag was lost two days, then given back to her with the unmistakable tread marks of having been run over and crushed by a ramp baggage tractor. The commuter line used American Airlines' gates at DFW.

Diane was furious that this war among giants had come home to her personally. She raised hell and knew how to do it. American Airlines finally coughed up a check to her for $39. It was a $196 Skyway. Diane is still too mad about this to ever cash their check.

There is no doubt that in some cases there was bad blood between Braniff and American, who quickly spread out into the hard-to-get DFW gate spaces vacated by Braniff. Gate spaces at a big international terminal are harder to get than airplanes.

Fat Albert, now humming its way over the Pacific, had already become a legendary airplane. Magazine and trade articles had done stories about N601BN as the best utilization of a 747. This was in a day when there was some question about the need for an airplane that would dump 400-plus passengers at one time at any airport anywhere, much less the unlikely idea of Texas to Hawaii. But Fat Albert never lost money. Boeing complimented Braniff on their progressive maintenance program and utilization of their 747.

The giant left Dallas as Flight 501 was refueled, got a new crew, headed back to Texas an hour later as Flight 502, before it could cool off. Depending on the time of year and load, the trip was usually 7 hours and 45 minutes outbound, 7 hours back home. Fat Albert did hustle. The plane received needed maintenance at the Dallas end and was mostly able to keep up this trans-Pacific shuttle seven days a week. There was a ramp worker at Honolulu who said to one of Fat Albert's crew, "Say, do all your 747s have the same number, 601?" No way could he believe that he was seeing the same airplane and that it was the only one Braniff had then. Bright orange, white contrast and lettering, Fat Albert was hard to not see.

Only Braniff's most senior crews were able to bid for and hold the Honolulu run. The cabin attendants, a crew of 17, kept house in Fat Albert like it was their own. Not only was the food and service outstanding, but the huge cabin was Dutch-housewife clean. Charlie Brown told Diane they were working on a third cabin annunciator sign to go with "Fasten Seat Belts" and "No Smoking." This one would read, "Keep Your Gahdamn Feet Off The Seats."

There is another Fat Albert legend that received little notice in the press, but typified the Braniff family. A captain suffered a swift and fatal heart attack, inbound to Texas and past the point of no return over the Pacific. The Captain's wife was one of the senior hostesses working in the aft cabin. Married couples bidding for and flying the same flight were not uncommon on Braniff.

First, the forward crew asked their dozen or so upper-lounge passengers if they would move down to the first-class cabin directly below, and told them why. Then they carried the stricken captain up to the lounge where they could lay him out on the deck. They found three physicians aboard who tried every means to revive the skipper, all to no avail. Then they made the decision not to tell his wife. Why have her grieve the rest of the trip?

A second decision made by the crew was to continue the flight on into DFW, although the 747 was not certified to be flown by a two-man crew. Their reason? To land at Los Angeles would strand the grieving wife with her husband's body on the west coast, leaving her among strangers and facing the long, costly arrangements to get back home to Dallas. Furthermore, the delay of the flight while waiting for a relief captain to be found would scramble the Dallas connections of every soul on board. So they flew it right over LAX and home to Texas—a homebound 747 with a skipper who died suddenly of a heart attack, still wearing his hard-won wings and four stripes. Nobody got cited by the FAA, and the wife-hostess was surrounded by the support of friends when, on the ground back home, she learned of her husband's death.

Now, on the final flight of Fat Albert to Honolulu, the layover crew, waiting to bring the airliner back, did not expect to see it arrive. Charlie Brown said she was not only wondering how she'd get back to Dallas, but what in the world she would do with all the junk she had accumulated at the Honolulu end of the trip. Airline crews usually travel light and pack tight, unless they get into a pattern where they know they will be staying awhile at the other end of the run, in which case they gradually move the comforts of home into a distant hotel.

Airlines rent a block of rooms at turnaround-point hotels and the hotel becomes another family center. In Honolulu, as the word swiftly spread that they were now unemployed and stranded, the pilots and cabin attendants all drifted down to the old familiar coffee shop, needing to be with each other, trying to solve their monstrous logistics problem.

Charlie Brown said, "Some of us were already moving stuff out of the rooms and into the lobby. A monster pile was gathering."

Then the hotel manager came in, looked them all over sadly, and said that, in view of the fact that he would undoubtedly be stuck with the company's hotel bill, and that he had come to know and like so many of them—sitting there in his coffee shop, staring at him—the least he could do was NC (no charge) today's coffee. There were cheers and tears for the manager.

Then, as if by a miracle, word spread that Fat Albert was at Honolulu International and everybody was to haul ass, bag and baggage, and get on board for the turnaround trip home.

Those who were working the trip and not yet in uniform made haste to get right, and the almost-stranded Braniff family made the sorrowful trip out to the airport carrying all their earthly goods like a stream of refugees. Which they were. Fat Albert had room for them and all that they carried. The evacuation of Honolulu was total and complete.

Captain Gordon Winfield, who was then Braniff seniority number one, gave them a quick briefing. "Let's don't make a funeral trip out of this. No crying. Same great service." Winfield had ordered the newspapers kept off the plane, somehow they had paid cash for the fuel, and Honolulu International Airport gave them the same fine ramp service, knowing full well it was only a last gift.

Charlie Brown leaned over our kitchen table, her face tight with grief, telling us the rest of the trip. "We served 'em lots of booze going home. We had no newspapers but they didn't ask, they knew. The worst of it was nobody spoke of it, but a passenger would make eye contact, pat my hand. There was no way I could keep from crying."

"I stayed busy, but kept thinking, this is the last time I'll cuss this slow oven . . . last time I'll put on this apron . . . things like that. We touched down at DFW at 6:00 A.M., right on time, but it was still dark and raining, sheeting down the sides of the airplane."

"I put the ribbons on the doors, set them in manual, all the routine stuff. We went way down to Gate 24 instead of our usual dock. The most gut-cutting sight was that sea of 727s in Braniff colors over there through the rain. Rows of them, in fours, dark and silent. There were no lights on at that end of the field. I got off first, with three unescorted children to help them find their dad. Reporters pounced on me. I just pointed to my little ones and kept going. Down the hall inside the terminal, *our* hall, it was all dark. Locked up. The secretaries had been allowed to carry out their middle desk drawers and potted plants yesterday and now nobody could go back in. There was nothing to do but find Bill [her husband, then a Braniff pilot] and go on home . . . fifteen years . . .''

Charlie Brown sat up straight and raised her beautiful tear-streaked face defiantly, "We were *good*! Nobody had cabin service like ours!" Her fist pounded the table top. "We had 'blitzing.' It's just our word." Then she leaned forward as if to tell me one more secret in addition to Charlie Lamb being a hero forever.

" I know you never heard of 'Doc' Speigle. He was just one of the best pilots we ever had. I don't remember how long ago he retired . . . but after Charlie Lamb set it down in Honolulu he took the aircraft logbook to sign off the trip. He signed it 'Doc' Speigle. And then the co-pilot and the engineer, they all signed it the same way. All the last signatures in the last flight of Braniff's last airplane, old 601, are signed the same way, 'Doc' Speigle. On that last flight of a great airplane for a great company that is now just a memory, the logbook has nothing in it but the name of a great pilot who is just a memory too, as captain, co-pilot, engineer . . .''

She stopped. She was weeping.

20

Sea Stories

IN MY OPINION there is no equal to the flying skills and personal courage of the Naval aviator. And from the who-splashed-who reports of the war in the Mediterranean Sea, one could say the same of Naval aircraft.

Admittedly, there is some prejudice here. No attempt to be fair and equal. I am ex-Naval personnel, World War II vintage, a Shellback, and a Plank Owner. That means I crossed the equator, and was aboard a ship when it was first in construction and commissioned. Of the few things I am certain of in my life, there is a right way, a wrong way, and the Navy way. And no matter how bad things ever get, I take comfort in remembering, "Far worse things have happened at sea."

Over the years since my honorable separation from the U.S. Navy, I have kept in touch. Mostly by subscriptions to two fine magazines: U.S. Naval Institute *PROCEEDINGS*, an overview of the Navy's present thinking, ships, and men; and *APPRROACH*, a colorful and delightful magazine which deals only with Naval aviation, particularly in matters of safety.

Sometimes the amateur writers, officers in the fleet, send in jewels of writing that could never be improved upon, such as this one from LCDR S.K. Gryde, VP-65, describing in-flight ice.

> With ice or frost on your wings, you are now the pilot of a new, unique airplane with a wing design that has never been tested.

Isn't that beautiful?

In my hurricane hunting days here on the Gulf Coast I reported from the eye of nine hurricanes. Two by air, one of these as solo pilot of a wracked and leaking Cessna 182, one as reporter aboard a Navy Super Connie complete with all those radar-dome bulges above and below. We tracked and entered into Hurricane Carla in that one.

Before departure from NAS Jacksonville, I was asking background questions. Why did the Navy use the Super Connie for this work?

"Well, we started out using the Navy version of the B-24 after the war, then began to fly Lockheed Hudsons into the storms, then began using Lockheed Neptunes, then Connies."

"Can you tell me why, sir?"

"Well, the Connies are the only ones we ever got back."

I had forgotten the wonderful way the Navy deals with questioning landlubbers.

In his crew briefing, the pilot was advising us of emergency procedures of the heavily laden craft on takeoff. "There will be three rings of the alarm bell, followed immediately by the co-pilot and myself."

In the actual storm penetration we flew VFR in the rain-soaked 800 feet of clear air beneath the hurricane. The pilot handled the yoke, the co-pilot the rudders. Both were soon soaked with sweat. Somewhere over the central Gulf of Mexico the flight engineer leaned forward and casually reported, "Lost oil pressure, number three."

"Feather it," said the skipper, and out of my porthole I saw the blade stop and point edge-into-the-wind.

The old chief of the airplane, who saw me looking, and the pale color of my gills, said, "What will you do if we lose another engine?"

"I already done it".

I was more or less accepted from then on.

We returned to JAX and picked up another Connie. We made it out to the eye of Carla in this one. The storm was so huge that we could do a one-needle climbing turn inside the eye; we broke out into CAVU on top at only about 10,000. The hurricane covered the entire gulf, but was that shallow.

We circled the eye, making various weather observations and sending them in to the Miami hurricane center. Atmospheric conditions are so bad in a hurricane center. Atmospheric conditions are so bad in a hurricane that only the old low-frequency Morse code penetrates. With copper wire from tips to rudders, and an old time lightning slinger bent over his Morse key, we sent data that evacuated whole towns. Including my own.

Circling the eye was beautiful, an angel-food layer on top, gradually turning to darker shades of blue on down to the wave-lashed sea below. A *LIFE* photographer aboard wanted to make that a cover picture, if he could get it. Would they open the back door, tie a rope around him, and hang him out in the storm? They did.

It looked like such a great experience I asked them if they would hang me out too. They did, and I still have the breathtaking photos.

Later in life, the Navy toured their training carrier, the *Lexington* to gulf coast cities and brought reporters aboard for an overnight stay.

They used a little radial-engine Grumman twin for what they call "C.O.D." flight, Carrier Onboard Delivery. With backward-facing seats, which is a good idea gone unused by civil aviation, and with not much view under the high-wing engine nacelle, a first carrier landing is more like a crash than anything else.

The routine landing of a jet fighter aboard a carrier, night or day, is more adventure than the average shore pilot can accumulate in a lifetime of flying. Let me tell you why.

First, the Naval aviator must be able to locate his airport, which has been wandering around here and there at up to 30 knots while he has been gone. There are no landmarks on the ocean. It all looks the same. Then the pilot must fly a centerline approach to a runway that is too short for his high-speed landing and is canted away from the direction the ship is moving.

All his landings must be perfect spot-landings on a threshold that is heaving up and down twenty or thirty feet or so, and also twisting from side to side. He "acquires the ball," a red-light projection from the stern of the carrier on final approach, but close-in he must commit the great act of faith of giving his landing over to the LSO, called "Paddles" in the movies. A Naval aviator himself, the LSO talks and hand-signals last-minute refinements to this approach. A wave-off means go 'round, no matter what the pilot's opinion may be in the matter. The LSO stands exposed on the aft portside corner of the flight deck. It's about 60 feet down to the ocean.

The pilot has his wheels down and his tail hook down, his hope is to snare one of the cables stretched across, and slightly above, the deck. Not so high as to snare the landing gear, but high enough to catch the hook, and unbelievably chancy business.

So random is the chance of catching a wire that as soon as the pilot hears his wheels thud into the steel deck, he opens the throttle to full power. The simple reason for that is, if he missed all the wires, the hook may have hit the deck between wires and bounced high. If he "bolters," then there is a hope that he will still have enough speed, enough thrust spooled up, that he can fall off the forward edge of the canted deck and regain flight. Instead of going to the hospital as most of us would like to do after something like that, the pilot re-enters the pattern for another go at it. Snagging the fourth and last wire will stop a jet right at the edge of the deck, the pilot already out over the ocean blue. They volunteer for this. They do it at night, too.

Takeoff is a simple matter of being fired down a steam catapult at the speed of a cannonball. Most dreaded by Naval aviators is a "cold cat shot" at night. The pilot who must be ready for full IFR at the instant of launch, must also be able to detect if the explosion was not full force and the airplane is not going to fly. He decides this in the length of time it takes to drop the airplane overboard into the sea.

With instrument reference only, he must decide whether to punch out of a few million dollars worth of the taxpayers' money or be forever removed from his family's gene pool. The Navy fighters, like all of our military fighters, use a zero-level ejection seat which is good unless the falling airplane has rolled and the pilot ejects into the side of the carrier.

Equally honorable in my opinion are the crews of the long-range Naval air patrol craft. These folks, while not getting the everyday thrills of carrier operations, fly endurance-contest missions that carry them routinely across the broad back of the blue oceans to places where not even God can find them.

The current Navy long-range patrol craft, which in civilian clothes was the old Lockheed Electra, is the most dreaded enemy of an unfriendly submarine.

Here is CDR T.A. Richardson's story of an encounter with a Cessna 150, as told in the pages of *APPROACH*.

It was the first light of day and the big turboprop P-3 was in the pattern of NAS Southwest, verifying some maintenance work on a squawk about asymmetrical flap deployment. The flaps were now working in pairs.

On final the PPC (Patrol Plan Commander) saw what he first thought was a spot of water on the runway ahead. But there had been no rain there in over a week, and the spot of water appeared to be moving.

The PPC checked with the tower, who replied, "No traffic." But by now they were close enough to make out the details in the dim early light and decided the "water spot" was a Cessna 150. The PPC cleaned up and initiated a go'round. There was, indeed, a Cessna 150 on the runway.

It seems that the Cessna was being flown by a student on his first cross-country. He had mistaken the Naval Air Station for his destination airport. Because the field that the student thought he was on had no tower, the student had not bothered to make a radio call or turn any lights on. He had slipped by the tower neatly and unseen, in between the touch-and-goes of the P-3.

Now, there was much attention being paid to the little 150 who had turned off the active and was attempting to taxi upstream into the whirling blades of a whole squadron of P-3s, which were moving off the ramp to start their day's operations.

Still unable to raise the Cessna on the radio, and seeing it in danger of being diced by oncoming propjets, NAS Operations quickly surrounded the Cessna with a circle of yellow trucks and herded it off the runway toward base operations.

Here I must quote CDR Richardson and his tight language in *APPROACH*.

> After the predictable confrontation with the OPS officer, the distraught student pilot climbed into his steed, started, taxied, and took off just as he had arrived—covertly, much to the agony of the NAS OPS yellow trucks chasing him. Good stick. Not much of a talker.

All of which brings to mind one fine Sunday. I was homeward bound through Pensacola, and there lay the carrier *Lexington*, moored dead into the wind. The devil jumped up and sat cross-legged in my lap. With 40 degrees of flaps and enough wind off the Gulf, I was sure I could put that little Cessna 150 on deck and stop in time. I could have claimed carburetor ice. No evidence. The only thing that worried me were those cables. Were they retracted down flat on the deck, or were they raised? In the raised position they would be high enough to trip that 150's wheels sure as heck.

So I flew on, still filled with respect and admiration for the Navy, but wondering if one landing on the deck of the *Lex* would have qualified me for carrier pilot.

21

About Spins

ABOUT 30 YEARS AGO when I learned to fly, airplanes were simple, aviators were simple, and the FAA was only some distant government agency that I had heard about now and then. Around Beaumont, Texas, we had no such things as flight schools. All the prospective student had to do was try to find an airplane and somebody who would give him a little dual time in it. Dual instruction cost $10 an hour then, and I have a few half-hour entries in my early logbook, which meant I only had five dollars and had spent it flying.

This random pattern of finding an airplane and instructor—at the same time and place—gave me a healthy variety of airports and airplanes. Champs, Cubs, Luscombes. I could fly with either hand, but couldn't anybody? Sometimes I could find an instructor at Beaumont, sometimes over at Orange. We had three airports, two of them grass.

I got airsick at first and carried a little bucket with me. I really wanted to fly, and I soon found out I didn't get airsick if I was the pilot, unless we were doing spins. We all had to know how to do spins. They were exciting, but I dreaded them. Not from fear, but because the earth whirling below like that would really make me sick in a hurry.

I learned that an airplane which will spin easily will recover easily. That Luscombe was just a spin waiting to happen, but if you stuck its nose down and turned it loose it would stop spinning instantly by itself.

Somebody wrecked the Luscombe just before my private pilot check ride, so I flew a Cessna 172 which was brand new and one of the first ones I had ever seen. Square rudder, polished aluminum, it looked as big as a DC-3 to me after climbing in and out of those little two-seater rag dolls. I had never flown an airplane with a nosewheel, radio, flaps, or a trim wheel. But I loved that 172 from the first, and I still do. I flew a fairly good ride in it, I guess. He passed me.

Of course, none of us used the radio much then. I could get in or out of Houston Hobby with just a red light or a green light and a friendly waggle of my wings. I wonder if we could go back to that. Today I can't get into Hobby with three radios and a transponder, all of which are worth more than the whole airplane I flew back then.

After I answered my 50 questions and got my license, any additional flying knowledge was just stuff I picked up by hearsay while hanging around the airports. And at the same time I learned a lot of stuff I shouldn't be doing but didn't know the difference.

We had a 172 pilot who claimed he could loop his plane with a glass of water on top of the instrument panel and never spill a drop. So I thought it must be okay to loop a Cessna. Oh sure, there were some "category prohibitions" printed on worn placards fixed to the cabin side, but I thought they might have been some kind of advertisement.

While spins scared me and made me sick, there is nothing so beautiful and pleasurable as diving an airplane until it sounds serious enough, then hauling back on the yoke and rushing up, up, and over, pulling the earth into view upside down at the top of the windshield, tightening up the loop a little if the controls felt wishy-washy, then riding the great swoop down to level flight right where I had been before. Sometimes I'd do this on cross-country flights just to get wide awake again. Nobody ever showed me how to do a loop. It just looked easy, and it was. Listening to the plane, flying by feel (I never looked at the instruments), and coming off on the power on the back side so as not to strain anything.

Then one day we got one of those little skinny-backed Cessna 152s. It was brand new. I took it up and played around awhile, slow flight, stalls, a really fine little airplane to fly. I decided to loop it.

About Spins 99

Both cabin doors came open at the same time at the bottom of the loop.

I said, "Lord, this is Gordon."

"I know."

"Lord, if you will let me get this skinny-backed airplane down all in one piece I promise not to do any stuff like this again, and to start reading FAA books and going to church on Sunday."

He didn't answer, but I got it all down in one piece and never told anybody until now. And I kept most of my promises.

Soon we had become Civil Aviation. It took a radio even to get into Beaumont. And I became a Civil Aviator. But the lust was still there.

Just a little lust, understand. No outside loops or inverted spins. I just loved to gracefully raise my wings and roll the earth and sky about. In my heart I could hear waltztime. But I had gotten checked out in the Stearman, the beautiful two-holer that belonged to George Mitchell of M&M Air Service, an airplane whose song I have sung many times before.

I liked lazy eights and back-to-back wingovers. I wish I had learned to slow-roll it and barrel-roll it, but that took feet too, and my feet kept floating up under the instrument panel. No matter, I wouldn't have known what to do with them anyhow. I never learned to do these things. I was afraid of getting airsick in front of an instructor, and I wasn't good enough to teach myself.

When I was doing aerobatics in the old biplane, I would go up to about 4,000, which takes a long time in a Stearman, and spin it first. I didn't like the spin, but since so many of my attempted maneuvers ended in a spin anyway, I wanted to do one, get it over with, and get used to it.

I could do a fairly good snap-roll, but nobody ever told me how to un-snap it. So the airplane would groan and whoosh through a snap, then try to do the next one because I hadn't moved the controls, and that always ended in a spin. Sometimes I would get into a spin from a nose-high attitude and have no horizontal reference and not much idea of what the big old biplane was doing. That's when I began to love and trust the Stearman for more than just its good looks. I would just turn it all loose, retard the throttle, and the heavy end always came down first.

There were some funny things about spinning a Stearman. If you didn't punch into a spin just right, the airplane would not spin but, instead, enter into a tight flat spiral that you could just fly out of.

But if you got the right combination to make it spin, oh boy! At about the third turn the Stearman tucks her nose straight down, and the cockpit coaming feels like it has closed in and grabbed you under the arms, "Gotcha!" The old dog really spins. With a loud whooshing of wings and wires it makes a pinwheel of the earth below. It takes a bit longer to come out of one of those, and that scares me bad.

Being a Civil Aviator now, and with lots of sweet ratings and flying in a Civil manner, I don't even pursue stalls past the recognition point. There is really no reason why one should.

And so it came to pass one day not so long ago that I was checking this Civilian out in our Mooney. No, I have never done anything fancy in that airplane, although I know it is strong enough and its lines look like it's begging for a workout. The most I ever do in the Mooney are paired wing-overs on a long cross-country, just to break the monotony. Laughing with glee as we come storming out of the second one, we pick up original heading and altitude.

I have always wondered if Center can see you do a wing-over on radar? Anything that good must be illegal. Someday a white FAA station wagon will be waiting for me at the ramp: two guys in white shirts and narrow ties . . . with handcuffs ready.

But I do insist a check-out is not real unless we have done some stalls. So me and this checkee did a 360-degree clearing turn at 4,000 feet out in the practice area, and he showed me a power-off stall.

I thought he went into it too deeply, but the old '68 Ranger 20-C does break clean and level, bobs quickly, and flies again quickly. Unlike the wonderful old fat-winged Cessnas, which claw and flutter at the last rags of lift, the laminar-flow wing of a Mooney is either flying, or else the laminar flow breaks loose all at once and the airplane is not flying. But it goes back into business as soon as you get the nose down.

My check rider was making me uncomfortable in his power-on stall, but I don't like to meddle too soon and not learn what he knows. So we were all laid back in the seats, horizon lost, full power. He added more and more rudder to offset the tendency of the airplane to try to walk around the compass, and more and more aileron and back pressure on the yoke. But not for long. Just as I realized we were ideally set up for a snap, it snapped.

I mean, it went right over the top. The spin was born fully developed. I did not have to wrestle the controls from him; we both did exactly the same thing at the same instant. Yoke shoved to the panel, ailerons level, hard rudder against the spin.

Nothing happened . . . for what seemed like 30 years. Actually, we only did three quarters of a turn, but we used up about a thousand feet, falling out of control. After that the recovery was normal.

After we landed and got dried off I called a friend who is a test pilot at the factory and described what we had done and asked what they do for spin recovery.

"Yoke forward all the way, briskly, neutral ailerons, rudder against the spin. And yes, it does take about three quarters of a turn before anything happens, sometimes more. You did all the right stuff, as we say in the trade, but we don't do intentional spins here."

Not able to go quickly to sleep that night with the memory of the Mooney falling, I recalled a letter from a reader who told me of his pilot-ancestor who spun a Jenny back in the days when spins were most often fatal. They didn't know how to recover from one. This doomed soul in the Jenny faced death in anger. He decided to just open the throttle, stick the nose down and "get it over with faster." When he did that the Jenny recovered from the spin.

Or of the airmail pilots back in the twenties flying those big old DH-4s. Instrument flying had not been invented yet, and when one of these old boys got caught on top in fog, they would fly to where the sky-glow indicated their destination below, kick it into a spin, and spin down through the clouds on purpose, a controlled rate of descent. They must have had big ones. Made of solid brass.

Most pilots today learned to fly after spin training was condemned. Spin training was stopped because it was judged too dangerous. Also, the idea of spins scared off a lot of money, er, students. So we have a whole generation of pilots completely innocent of spins and spin recovery. That must be dangerous, too.

A few pilots enroll in some limited aerobatic training to learn such basics as how to get right-side-up again if wake turbulence or weather should upset them, and they learn spin recovery.

At the very least a pilot should read the aircraft operational manual to see what it says about how to recover from a spin. If the subject is not mentioned, and you are going to be flying that type of airplane very much, then call the factory and find out. Spins are sure not hard to identify if one has got you, and it's a great comfort to know where and how to place the controls for a prompt recovery. On some, like the little Cessnas, you just turn it loose. They don't like to spin and will recover by themselves. How nice.

All of us have had stalls and stall recognition. If one wing suddenly drops down from a deep stall, your flying machine is on the threshold of a spin. Another common spin situation is in the landing pattern turning from downwind to base. If the pilot decides to tighten up this turn because of wind drift or just a plain overshoot, he will be looking at the runway through the top of his windshield, hauling back on the yoke to tighten the turn, and using top rudder to keep the nose up. Already slowed down, he's ready to snap right over into a spin.

Don't worry about recovery from a landing pattern spin. Not enough room.

22

Barnstorming at Port Arthur

AS A BOY GROWING UP in the twilight of barnstorming, I was convinced that flying was a virtue and that only the virtuous could fly. These holy standards, conjured up in my ten-year-old mind, applied equally to both men and machines.

Like my morals, my world was confined. Little flying was taking place in 1933, least of all in the skies over Port Arthur, Texas. Information about aviation was scarce. All the aviation books were pulp-paper dime novels about the great men and machines (fictional or otherwise) of World War I, which had only ended 15 years earlier. Many of these men and airplanes were still in service. Books and movies about aviation were scarce but were rich in the details of flying in the Great War. Some lads of my age had minds full of baseball heroes and batting averages. I could recite, without faltering, the engines, armament, span, and speed of the principal warplanes of both sides.

Likewise, I knew the names of the aces who flew them to glory. Lean men with steely grey eyes, all of them. Men who may have bared their teeth or screamed during attack. Some were more than a little daft. Frank Luke, the balloon buster, for example. I knew from the way he fought and flew his SPAD that his days were numbered. Lt. Jimmy Doolittle, by contrast, was winning air races after winning the war. I could tell just by reading about him that Doolittle's star had only begun to climb and that he would probably live forever.

So complete was my romance with the sky that I turned in school papers with Fokkers, SPADS, S.E.5s, and Curtiss Hawks alive on the margins. Other than making a mess of my work, they only served to flag my papers as the work of an idle daydreamer.

One teacher was good enough, kind enough, to write an appropriate comment on my already crowded margins. With an arrow drawn toward a Curtiss Hawk P-6E, he commented, "If you don't settle down and learn some of this, you will never sit in the cockpit of one of these." I might have believed him had I known he was a pilot, flew a Fleet biplane with a barking five-cylinder Kinner radial, and liked to do spins for fun. He just didn't look like a pilot. Medium stature, a little smile, a mild way of standing, glasses, and the beginnings of a belly. None of my pilots looked like that, and none of them taught school.

Many years later, after World War II, this teacher and I met again. He looked the same, but I was much older. He asked me to come fly with him; he now owned a nimble and swift little Piper Pacer, wheel pants and all. He would fly out over the pastures and glide the white little airplane silently down, then open the throttle and stampede the cows, shouting and gunning his engine. A complete transformation of the man I thought I knew.

"Gordon," he said to me, as serious as he had ever been, "don't you ever tell a soul about this." He was sweating and laughing, and there was a light in his eyes I had never seen. He looked like a pilot.

Well, it's been 40 years, and that's as long as I can keep the story quiet. By now I have lived long enough to expect to find aces in different places.

Back in the thirties if we heard an airplane engine in the sky, all work stopped and we all came out to look upwards and see the airplane. One summer day while I was still in school the Navy routed either the *Akron* or the *Macon* over Port Arthur. School was out; we poured out onto the playground, searching skyward for such a heavy drone of engines.

"There she is!" All eyes looked toward the pointed arm. The mighty shape was emerging from a cloud, looking like a silver cloud herself.

For a moment we saw her entire length. Then, just as solemnly, she merged into a summer cloud again. That one silver nail of history bent over in my mind when I read that all the great ships in the sky had been lost and were gone.

We seldom heard an airplane engine over Port Arthur. We had what was called the "Texaco airport," but then, as now, Texaco did not encourage familiarity with the rabble. The field lay adjoining the great Texaco refinery, Port Arthur's sole reason for being, and we heard rumors that the great Frank Hawks zipped in and out in his famous Travelair Mystery Ship, Texaco 13. But none of us ever saw even a shadow of that.

My dad spent a lifetime with Texaco and once took me out to view the sacred grounds of the Texaco airport. From the marsh road rimming the field, with the refinery beyond, I could see a tin shack and the windsock. My mind could only visualize the lean, bronzed Hawks wrestling the Mystery Ship to earth here, standing briefly in the sun.

Then, arriving in quick succession in 1933 we had a barnstorming Ford Tri-Motor and Clarence Chamberlin with his Curtiss Condor. My parents bought me a five-dollar ride in both of them, and that was when my mother could buy a week's groceries at the company store for a five-dollar coupon book. When you scale the price of those rides to the price of today's groceries and what it cost to ride in the Concorde when it barnstormed Oshkosh in 1985, it comes out about the same.

More people rode in the Ford because it looked more modern. It had wicker seats and made a tremendous din. What stuck in my mind was opening up the door to the toilet. There was just a hole in the floor. Take that EPA and farmers of Io-way.

The Condor was my first airplane ride, and I had a little camera with me. I still have the pictures: standing by the propeller of that huge Curtiss Conquerer engine, wearing my oil-cloth aviator's helmet and celluloid goggles. An ace.

Not until later years did I realize mine was the first Condor model, same as the bomber. It had a box tail. Its V-12s were sleekly cowled but with a round radiator that stood on top like a small oil drum on end. Later Condors went to radial engines and an improved bracing design that left out half the number of interplane struts and all the charm. They had just an ordinary-looking tail assembly—one rudder, one horizontal tailplane. Most had modern-looking cowled radial engines and wings of unequal span.

I just couldn't get enough of looking at the Condor. It had just a vestige of a nose below the windshield, like the brow of a battle-weary tomcat. The exhaust stubs of its V-12 engines stuck out only a little. It ran with a slick roar, like a pair of P-51s going over. The passengers were protected from all these great aviation sounds—engines, propellers, struts and wires in song—by a layer of carefully painted fine cotton cloth. It had a dark blue fuselage and orange wings. I never saw a picture of a Condor painted differently.

I loved riding in his aeroplane, and Chamberlin was ever so smooth and gentle with it. Both of theses multiengine planes landed in a pasture near the highway. I don't know if they even asked Texaco.

At the end of this day's heart-stopping adventure, as everybody went home, I took one last snap of forlorn-looking Chamberlin standing alone in the door of his great airplane. He was small and balding, and looked kind of sad. How would you feel if you had been the *second* man to fly the Atlantic. But he was nice and talked to kids.

I walked with a different swagger now. I had actually been up in an airplane. Twice. But walk I did, about a mile to school and back. The depression still lay heavy upon the land. We lived in a rented house out beyond the edge of town, just a shell road, fields, and meadowlarks for company as I walked home from school each day.

It was on such a peaceful daydreaming walk home one sunny afternoon, out through the fields, nary a tree nor a house, that a shadow fell over me. Only a coastal cloud, moving in from the Gulf, then a voice called from the sky "Hey Red! Do you know which way to the Texaco airport?"

I just knew God had cast his shadow over me and had spoken.

Before casting myself prone upon the earth I glanced upwards. God, if that's who it was, was leaning out the gondola windowsill of the Goodyear blimp *Wingfoot Express*. It had drifted silently directly over me.

Of all the pedestrians in Port Arthur whom the crew could have asked directions, few could have known more exactly, understood more completely, been more everlastingly honored than I.

"Yes," I shouted upward, while drawing myself to rigid attention—the proper stance for conversation with aircraft in flight. Then, slowly but without faltering, I began to rotate until I was facing the direction of the tall Texaco stacks, visible from all the way across town, a heading of about 280 degrees. I raised one arm, pointing and frozen in position, exactly to where the distant airport was.

The blimp stirred itself to life. First one engine barked, then the other. Slowly the aircraft pivoted overhead and began moving down the arrow of my pointed arm. "Thanks, Red," came the faint shout from above.

With solemn dignity I drew the extended arm up into a salute. The blimp wagged its rudder at me. It was a moment that I hoped would live forever.

Aviation moved toward the pastures of our end of town, away from the refineries. We knew every field. A strange pilot circled a strange airplane over a field that we knew had a hidden bootlegger's pit in the middle of it, masked by tall weeds. We had bicycles by then, sped to the pit, stood beside it, and made frantic wave-off signals. The pilot flew by, low and slow, staring blankly through his goggles. One of us stood down in the pit to illustrate what we meant. The pilot nodded, gunned his engine, circled the field, came back, and landed on a safe smooth side of it.

The plane was an OX-5 Lincoln Beachey, not a common airplane, then or now. The pilot gave us a free ride. He was lean and brown, and carried a six-shooter tucked down in his rear cockpit. Many years later, reading of pioneer aviators, I learned of only one smuggler operating on the Texas Gulf Coast with a Lincoln Beachey. That surely was him.

In the late thirties, one straight-wing J-5 Waco kept coming back each summer. It was Pop Johnson and his two sons. He flew the biplane; they followed in a cream-colored supercharged Cord convertible. The Johnsons worked the same field every summer. Me and my best friend Jim always got there first, held up a flag we had made to show the wind, and slept in the Waco every night to keep the cows from eating the fabric. We always got the first ride the next day.

I have since traced the airplane. It got flown to junk, cropdusting at Stuttgart, Arkansas. The pieces were bought by a Waco restorer in Florida. It lives again today, almost 50 years later.

The last barnstormer came in a dark-green cabin-C Waco. And he didn't fly away. He built a little shed, and then a hangar, out there on the edge of the highway where the Condor had flown. A quiet man, his name was Glen Parker. He bought a Cub and gave lessons.

Soon the seed of aviation sprouted at what was then called Parker's Airport.

Parker is gone, the place is a subdivision now, but lucky is the local who can claim Parker taught him to fly. He was of the first generation of pilots here.

Most of that generation have gone winging off into the sunset now, men who flew for the love of it, long before the airplane was ever called, by someone wearing a coat and tie, sitting at a desk in an air-conditioned office, a "useful tool."

23

Jeana's Eyes

I'M IN LOVE. It's hopeless, I know. She is spoken for, and I have a family back in Texas, but I knew it was going to happen. I fell in love with her pictures. That didn't worry me. I could just keep quiet about it and go on through my life, safe in knowing she would be far above me on silent wings and I would never meet her.

The only trouble is I just met her. In person. She is even more beautiful than any of her pictures. I am smitten. Not just a convention affair. I might as well come out with it. When I get home my wife will look me over for a minute and then ask, "All right, who is she?"

Her name is Jeana Yeager. She's an aviator, if that's any help. A much better aviator than I will ever be. She has set four speed records flying a Rutan Long EZ, and of course, she has been chosen to fly as co-pilot with Dick Rutan on the nonstop, round-the-world, history-making flight of the Voyager.

I think Dick Rutan, pilot-in-command of the Voyager and of the organization behind this record-seeking flight, is her boyfriend. I don't know this. Never read anything in the papers about it. And when I met Rutan I didn't have the guts to ask him. But I'll say this, if he ain't then he ought not to be trusted in any airplane after dark.

I met her at the Voyager booth in the center aisle of the south exhibit hall here, where you can meet her too. She is working there, along with just about everybody else in the Rutan family, plus a few old friends. They are selling genuine autographed pictures of her and Dick and the airplane for $5, Voyager buttons for a buck, caps for $10, a desk model of the airplane for $200, and so on and so on, trying to raise money for the flight.

That seemed unnatural to me, that Dick and Jeana, soon to be immortalized in aviation history, are now out in a booth, largely unrecognized, hustling away for a buck or two. I had pictured them as being in the presidential suite of a grand hotel, a limo waiting outside, fruit baskets being sent up to their room. I was sort of shocked, but admired them all the more when Rutan came into our press shack for Editor Nigel Moll's interview and spoke of the hard road for cash.

Rutan told it all. He told of foreign capital offered, "But this is an American flight, an American airplane, and that just didn't seem right to me." He also turned down big bucks from a tobacco company advertiser. Doesn't believe in tobacco. I liked him more and more. And like Lindbergh, he turned down another fortune rather than give up flying the airplane himself, the airplane that he, his brother, and his friends had designed and built at the Rutan factory. So Dick Rutan has his pride, his ethics. He also has his hat in hand, asking for a dollar here, a dollar there. Not many of us knew that.

He has formed a Voyager VIP membership for persons who give $100 or more. The forms to join are at the Voyager booth. As he spoke to Nigel and me of avoiding weather and all the other risks of this flight, I could also see the worry over money in his fine warrior's face. The great common problem of being just plain broke.

The one kind of money I know Dick and Jeana would accept with grace and heartfelt thanks would be from us, the EAA, fellow builders of experimental aircraft. They need something like $400,000. If all of the multitude of you here knew this and felt as strongly about it, Dick and Jeana could take that much money home with them from Oshkosh.

Jeana posed for a picture with me at the booth. Stood right beside me. I can still feel where her hand rested lightly against my back. And I will never forget her eyes. The color of the skies, they dance with some mysterious inner light, some enchantment, some excitement I have never seen before.

There is much to be said for ex-combat fighter pilot Dick Rutan's choice of Jeana as co-pilot. With the exception of Lindbergh, nearly all of the pilots of world-record-setting flights took some big awkward fellow along as co-pilot. Dick is not only taking one who is little and light, and an expert pilot herself, but on the lonely two-week journey over distant oceans she will be easy to look at.

I will follow the press reports of their epic journey a year or so from now, and as they round Cape Horn, I will still see Jeana's eyes.

24

The Ten Best

IN THE SUMMER OF 1985, *FLYING* asked six of its writers to list what we thought were the Ten Best airplanes. Limited for space, they asked us to keep it short. Here are mine:

****

To me, for an airplane to be among the Ten Best it must be honest and utterly reliable. No witchy, tricky designs. I believe that in airplanes, as in art, only the truth lives.

Good airplanes are ageless and often pick up affectionate nicknames. Jenny, Tin Goose, Gooney Bird, Herky Bird, Connie, to name but a few. And yet the 172 is just a 172, although it is one of the best airplanes ever made, and just as loved.

So what are my Ten Best? Airplanes faithful as a bird dog, kind as Santa Claus. There is no reason for the order of this list, only that I had to start somewhere.

1. Aeronca Champ
2. J-3 Cub
3. Cessna 150-152
4. Cessna 172-182
5. The Mooneys
6. Jenny
7. Stearman
8. DC-3
9. Lockheed C-130
10. Boeing 747

AERONCA CHAMP

The Champ is a little airplane without fault. This pot-bellied little daisy-cutter is pretty to look at, seats the student up front, takes less coordination than the nearly-as-good Taylorcraft, and flies more like a real airplane than a Cub.

The Champ will never sneak up and do something dirty to you, and the beefed-up version, called the Citabria (airbatic spelled backwards), teaches airbatics.

The Champ will not humiliate the novice taildragger pilot. And best of all there are still thousands of these honest little rag-wings still available in all conditions and all prices.

J-3 CUB

But isn't a Cub about the same as a Champ? Oh, no sir. The solo or student sits way back in this canvas bathtub; passenger or instructor sits between him and the instrument panel. The side opens—window up to the wing, lower panel down on the strut. Fields, farms, and traffic will slowly pass in the open view below.

The long swing of the fuselage, the unbelievable lift of the fat Cub wing—they open worlds yet unknown. If you can glide a Cub down to a no-bounce three-point landing, you have mastered pitch and power.

Pilots get out of Cubs full of laughter. Sorry, you waited too long to find a cheap one.

CESSNA 150-152

This two-place Cessna will do you no harm. How many, Lord, how many has this everlasting honest metal airplane introduced to flying and kept them there.

The baby Cessna won't let you cheat but won't scare you either. The old ones with 40-degree flaps could do carrier-deck landings.

The little Cessna will teach you to fly, yet is speedy enough to fly you far away in one day. Requiring little care, a best buy in used airplanes now, thousands are parked awaiting love and care, and there are no "bad" models to beware of.

CESSNA 172-182

The 172 is the best-selling airplane in the world, and rightly so, but the slightly larger 182 is a move up from Chevy to Cadillac. You can feel it all over. Both are roomy and good instrument platforms. A good 182 will keep up with an old-model Mooney. Legend has it that you can haul whatever you can get in the door, but don't do that. There are some 172 models with flaky engines to beware of. But there is no such thing as an old 172 or 182. A 172 pilot can make the transition up to the heavier-handling 182 about as quickly as he came up from 152 to 172, which is what Cessna had in mind all along.

THE MOONEYS

When this column appeared in 1985, the Mooney 231 was king of the roost. Since then the 252 has taken over as the airplane that flies fastest, highest, and for the least cost. Because all Mooneys are built on the same M-20 airframe certified by Al Mooney in 1955, and all of them have always been the fastest and most economical in their class, let's just give you a Mooney table to look at. They are all good, except for many ADs on the early wood-wing and wood-tailed models, whose owners are devoted to them.

1955 M-20: wood wing and tail; 165-hp Lycoming.

1958 M-20A: 180-hp Lycomings used until 1977.

1960 M-20B: metallized by Ralph Harmon.

1962 M-20C: a refined B-model; 180-hp Lycoming; a vintage model that stayed in production beside later, faster models.

1963 M-20D: limited run of "economy" models with fixed gear and prop; many converted back to C's.

1964 M-20E: optional 200-hp Lycoming. *(Also in 1964 the pressurized M-22 was built, but only 27 of them. Said to be the airplane that broke the company. Not an M-20.)*

1967 M-20F: 200-hp Lycoming and fuselage stretched 10 inches; good airplane. *(The M20-C stayed in production right beside it.)*

1971: *Butler International closed the doors after building a few M-20C's with the "button hook" tail.*

1973: *Republic steel bought the closed plant and retained the M20-C, -E, and -F models.*

1977 M-20J: the "201" refinement by Roy LoPresti; might be called the first "modern" Mooney. *(Mooney begins to name models after their top speed.)*

1979 M-20K: LoPresti lengthened nose 10 inches to fit in a 210-hp turbocharged Continental six. This is the "231"—as in mph.

1985: *New group, headed by Mr. Alex Couvelaire, Paris, buys company. LoPresti named the president. LoPresti's 301 test-flown successfully, shelved. Company brings out the* **M-20H**, *or the 252, breaks all existing speed and altitude records. Remains in production as Cessna and Beech suspend piston-engine production and Piper reduces to one model. Mooney backlogged with 252 production. The 201 finally replaces the M20-C as lowest-priced Mooney. (Our 1968 M20-C has retained the same resale value for ten years, but we can find no reason to sell it.)*

JENNY

One of the first tractor biplanes, one of the first to have an enclosure for crew. Powered by the faulty and unreliable Curtiss OX-5 engine of 90 hp, the Curtiss JN-4 or Jenny, served as ours and Canada's basic trainer during World War I. We built too many of them.

Cheap postwar Jennys flooded the market and suppressed the design and sales of more modern aircraft. But Jennys opened the skies of America to aviation. They barnstormed every field and farm; airstruck young men, such as Charles A. Lindbergh, bought them and taught themselves to fly. General aviation began with the Jenny. Made of wood and wire, most wore out quickly, but some lasted into the thirties. They were beautiful.

STEARMAN

The Jenny of World War II, officially the Boeing PT-13 or -17 (depending on engine) and named the Kaydet. But nobody ever called it anything but a Stearman, for Lloyd Stearman who designed it in the late thirties. Rugged and beautiful, the big biplane was simple to fly but considered safe on the ground only after it was tied to something.

Agricultural aviation began with Jennys then a few surplus Huff-

Daylands. But in the postwar years, and for the next quarter century, it was mostly Stearmans. The $200 surplus biplanes now go for a prized $20,000 and provide thrills at air shows. Any pilot who ever flew one remembers it with emotion.

DC-3

It is said that the last flyable DC-3 will appear sometime late in the 2000th century. Force landed into a lofty snowbank during an Alpine storm, the crew was rescued but left the cockpit sealed with ops manuals and logbooks inside. Snow and ice soon buried the "gooney bird," which is expected to eventually emerge unharmed at the foot of the glacier.

The DC-3 was a near-perfect design, the first to make money just carrying passengers, as they still do today, 50 years later. Flying a "3" is like dancing with your aunt who was fat but was a good dancer.

LOCKHEED C-130

Able to deliver supplies, troops, and trucks anyplace in the world, the C-130 is thought by some to be bulletproof. It is the cargo plane perfected. Ugly, some say, but beautiful when it's coming to rescue you. "Herky bird" pilots fly them with their fingertips. Once, one landed on an aircraft carrier, showing off. The C-130 is one of the best American things since the hot dog.

BOEING 747

I am not all that safe and comfy in some wide-bodied jets. In a 747 I sleep as if in mother's arms. Pilots have said it's an easy plane to fly. What else could carry the huge heavy space shuttle across this continent on its back?

The 747, the fully automated, all-electric-hydraulic American super airplane. There is nothing else like it.

****

If this leaves you wondering what planes the other editors at *FLYING* picked, here are their lists (with numerical notes as to how often the same plane was chosen):

Len Morgan

Bleriot XI
Sopwith Camel
Boeing P-12E
Fleet Finch
North American AT-6 Harvard

North American P-51 Mustang
Douglas DC-3 (2)
Douglas DC-7C
Boeing 707-320
Concorde

J. Mac McClellan

Boeing 707 (2)
Beechcraft Bonanza
Grumman Gulfstream
North American Sabreliner
Cessna 140

Sikorsky S-76
Concorde (2)
Lockheed P-38
Boeing 747 (2)
Cessna 310

Nigel Moll

Siai Marchetti SF 260
Concorde (2)
Pitts S-2B
Beech Bonanza (2)
Lake Renegade

Robin DR400
Cessna 172 (2)
Boeing 747 (3)
Vickers-Supermarine Spitfire
Hawker BAe Harrier

David Almy

Gates Learjet 28/29
Hughes 500E
Space Shuttle
Gulfstream I through IV (2)
Douglas DC-3 (3)

Cessna Caravan
Boeing 757
Piper Malibu
Cessna Citation S/1L
Beechcraft Bonanza (3)

Richard Collins, Editor in Chief

Bonanza 35 (4)
Douglas DC-3 (4)
Concorde (3)
Cessna P210
Beech 18

Gulfstream I (3)
Lear 23 (2)
Cessna 172 (3)
Piper J-3 Cub (2)

Among the six editors of *FLYING* there was a tie for first place: four votes each for the Bonanza and the DC-3. Airplanes getting three votes each were the Cessna 172, the Grumman Gulfstream series, and the Concorde. Both Collins and I voted for the airplanes we own—my Mooney, his Cessna P210. Len Morgan voted for the transports he'd flown. We are mostly what you might call a parochial bunch.

25

What Did You Do in the War, Dad?

I KNEW IT WAS GOING TO HAPPEN SOMEDAY. I even considered making up a lie about it. My little daughter Jenny, then "six and three quarters years old" as she said it, would have seen enough TV war movies and enough GI Joe, that she would come and ask me if I was in the war. And I would say yes. Then she would ask me what I did in the war.

There would be no use telling it all. I had been in the Navy the first half of the war, and then out, and those days nothing could be worse than walking around in your hometown in civvies and people wondering why you were not in the service. So I had re-upped and joined the Army Air Corps. That's what I had wanted to do the first time, but the requirements for being a cadet then included college, and I had never been to college. Now they were using up men faster, and it was easier to get into the flying cadets. I volunteered and was accepted for pilot training at Sheppard Field, Wichita Falls, Texas.

The way Jenny got around to asking me seemed simple enough at first. There were model airplanes hanging from the ceiling in her room. Sometimes she would ask me or her mother to come snuggle her down when sleep wouldn't come. As a crafty dodge she would ask me to tell her about the airplanes, knowing once she got me to talking about airplanes I would go on forever and she wouldn't have to go to sleep.

She already knew she was named after the Jenny hanging right over her bed, but tonight was the first time she had asked me about the others.

"Well, that little biplane, that's the one with two wings, is just a toy. But it's a Mexican toy. It was made in Mexico and it's all metal. It's not a real airplane of any kind, but when your uncle Frank Geirretts, the mad artist who studies in Mexico, saw it, he bought it and brought it to us as a gift."

"Well, I like it, whether it's a real airplane or not," said the little red head.

"The other biplane is a plastic kit model of the famous Jenny, as you already know. That is the airplane you were named after, only your mother thinks you were named after your great-great-grandmother Jenny. Both of us are happy with this arrangement, and it gives you some slack too. After you are grown up you can tell it either way you want to, and it will still be true." Jenny chuckled and nodded.

"Now, that silver plastic model with the long tapered wings is a DC-3."

"Was there ever a DC-1 or a DC-2?"

She knew that she had me going good and that this would last way past going-to-sleep time.

I told her there were lots of DC airplanes, that her mother used to fly as stewardess on the DC-6s and -7s, and for a little while on the DC-8s after we were married. And when mamma took her to Dallas to see Granny it was on a DC-9.

"What does DC mean? Is that in Washington?"

Smart little kid. I told her there was Washington, DC, which stood for District of Columbia, and with the airplanes it stood for Douglas Commercial. That they were built by a good man named Douglas, and Commercial meant it was for carrying freight and passengers.

"Was there anything after the DC-9?"

"Yes, but nothing to speak of at bedtime."

"Then what makes the DC-3 so special that you can get a model of it?"

I told her about the DC-3 being the first successful modern airliner, the first airplane that could pay its way by just hauling people, and that it was so good that all the airlines got started with them and some of the little ones, and airlines in foreign countries, still use them, even though they were invented about 50 years ago. And when I was a little boy before the war the first DC-3 came to Galveston, and we drove all the way over there just to see it. There were crowds of people from everywhere. They let me walk under it and all around it, and it was hard for me to believe that anything so big and shiny and smooth could fly. They even let me get in it and sit down.

"Did you get to fly in it?"

"Not that day. We didn't have the five dollars, but I sure did enjoy watching it take off and land. I dreamed I might someday be a DC-3 pilot."

"Did you ever get to be a DC-3 pilot?"

"Not a real one, you have to take special DC-3 schooling to do that, but I did get to sit beside the pilot and fly some from the co-pilot's seat."

"What was it like?"

"Wonderful. Then I thought of being a little boy in Galveston again, and I was filled with emotion."

"Was mommy ever a stewardess on a DC-3?"

"She's not that old. But she did serve on Convairs, which were supposed to be the replacement for the DC-3 but never quite made it. Your mother loved Convairs."

"Why don't we have a model of a Convair up there?"

"I can't find one."

"What made the Jenny so special?"

I looked at the clock, sighed, ahh-h well. I'd been waiting all her life for her to ask me these questions. Mommy was at school at a PTA meeting. It was just us. So why not?

I took a long breath, then began telling her that the Jenny was the primary training airplane for American and Canadian pilots during World War I. That the Jenny was the only airplane we got into production, and that we made too many of them. When the war ended anybody could buy a brand-new one, still in the crate, for just a few hundred dollars. There were thousands of them, and thousands of young men who wanted to be pilots, too. It was a wonderful time. These new pilots flew their Jennys all over the country, landing in fields and farms, and taking people for rides. That was how aviation got started in America, and that is why the Jenny is a special airplane. And I told her that, like the DC-3, the Jenny was a gentle and beloved airplane. But unlike the DC-3, they were not made of metal, but only of wood frames and cloth covering, so only a very few have lasted until this day.

"Are there any real Jennys left?"

"Only a precious few. With a little luck, someday we will see one. They are beautiful."

She just couldn't believe that an airplane could be covered with cloth. I told her how carefully it was stitched on, and made tight and strong by a special kind of paint called dope.

"Why does my model have yellow wings and a brown body?"

"Army colors. The yellow wings were so you could see it easily, the drab fuselage was so you couldn't."

"Dad, that doesn't make any sense."

"I know."

"I've seen pictures of you in front of a biplane with yellow wings. Was that a Jenny?

"No, that was a Stearman. The Stearman did the same thing in the second World war that the Jenny did in the first. It trained new pilots. I love the Stearman very much."

"Did you fly Stearmans?"

"Plenty. Still do. It's wonderful . . ."

Then my heart squeezed, uh-oh, here it comes . . .

"Were you a fighter pilot, Dad?"

This was it. To lie or not to lie. A little girl's vision of her father lay in the balance.

"No, honey, I was what they called a washed-out cadet."

"Oh. One night I heard you and some men laughing about being washed out cadets and you telling them about cleaning garbage cans. Is that what washing out a cadet means?"

"Well, in a way, yes. You see daughter, the Air Corps had a pilot for each fighter plane, but it took a whole lot of other men just to keep him flying. There were the guys who filled the gas tanks, guys who filled the ammunition cans for the machine guns, special men who worked on the engines and on the radios, and some who just patched up bullet holes . . ."

"—and my daddy washed out the garbage cans?"

This was going a lot worse than I thought. I tried to explain to her that an Army base was like a city, that there had to be men to do everything. And that they had big mess halls where all the men came to eat. Some of the soldiers, taking turns at it, had to do KP duty. KP was kitchen police.

"You carried a gun in the kitchen?"

"No, but I would have liked to."

"Did you kill anybody in the war, Dad?"

"No. I think the cook may have killed a few, but I killed none of them, and they killed none of me." I never thought someday I would regret it.

"Did you get wounded in the war, Dad?" This kid was relentless. Why don't they have TV heroes about buck privates doing what most soldiers did mostly?

"Yes. I was wounded in the war." I might as well get all this over and done with now.

"Oh, tell me how you got shot." She was wide awake.

"I was out on the gunnery range. After KP duty your dad was sent out to Kingman, Arizona, to be a goner on B-17s."

"What's a goner?"

"Same thing as a gunner. They had to teach us to fire the machine guns. The instructions began by putting machine-gun sights and handles on shotguns. Then a man in a hidden booth would fire little clay ashtrays at us which came over on the same curve as an enemy plane. There would be a whole row of us, standing and firing. Some of the kids, er, soldiers, had never fired a shotgun before, and the one standing next to me loaded his gun with it pointed level and pressed all the buttons and the trigger at the same time. His gun went off accidently and hit me with a load of number six bird-shot from a range of 50 feet."

"Oh wow! Where did it hit you?"

"Through my gas mask and into my behind."

There was a long silence. I could hear her trying to smother her giggles in her pillow. Then she just burst out with it, laughing hilariously, rolling over and throwing her arms around me. "Oh, my Daddy!"

Then all the rest of it.

"Did it hurt?"

"Plenty."

"What did they do?"

"Took me to the base hospital."

"Then what?"

"Nothing. Just painted me with iodine and left them in there. It was just birdshot, little bitty beebees. I still got 'em."

Uproarious laughter now.

"Did you get a medal, Dad?"

"They don't give medals for getting shot in the butt by the Americans."

She was suddenly asleep. At breakfast she told the story to her Mother, who objected to my use of the word butt. By noon it was all over the first grade. I always knew, even then, that not getting to be a fighter pilot was going to mark my life, mess me up with women.

"Dad, I think you're a great pilot now," she had whispered just before sudden sleep.

"Yer dern tootin' I am."

But I wasn't a fighter pilot in the war. The kid will just have to adjust to that. I did, almost.

26

All Chuck, No Yeager

FIGHTER PILOT Dohrman Crawford (Major, Louisiana Air National Guard) has a heart of gold. He had read my lament of how, when the Army Air Corps washed me out as a fighter pilot in 1943 and assigned me to KP duty, they probably prolonged the war by denying me my rightful seat in a P-40. For surely I was a natural-born fighter pilot, a nimble-handed, keen-eyed, hunter-killer fighter pilot.

I was sure back then that fighter pilots were a special breed of cat, and just as sure that I was born to be one of them.

Major Crawford had responded to this. He promptly wrote to say that my last chance to fly fighters was not yet gone. That I could come to New Orleans and fly his F-4 Phantom in combat against the Air Force all I wanted. He said to make haste if I wanted to fly a Phantom, as his squadron was being re-equipped with F-15 Eagles.

What's this? The Air National Guard flying top-line fighters? I had always pictured the Guard as once-a-month civilians wearing baggy fatigues, half-heartedly marching on weekends, shouldering M-1917 Enfield rifles. "Hup two three fo'. . ."

A little quick research revealed that much of our combat-ready air strength in both fighters and cargo planes is the Guard, that various units fly European missions with mid-Atlantic refueling as regular procedure, and that interservice competition for combat readiness is fierce. The Louisiana Air National Guard, of course, being one of the best (source: Louisiana Air National Guard).

The Guard's combat-experienced jocks, who are now bankers, lawyers, and captains of the line, claim that they have "more experience than pilots in the regular Air Force, but need haircuts."

"We get to fly our fighters as much as six days a month, same people flying the same airplane, serviced by the same ground crews who live here. We don't have to talk much on the radio, and we have a line of applicants a mile long, so we're able to hand-pick and choose our pilots and mechs."

An incident on the ramp at Eglin Air Force Base best depicts the relationship between the Louisiana Air Guard and the Air Force regulars. Just prior to a scramble, the Guard's old jungle-green F-4s were parked on the ramp nose-to-nose with the Air Force's new light-grey F-15s. On a softly spoken word of command, the Guard pilots left their planes, marched line-abreast across the short distance separating them from the Air Force aircraft, drew out their flight gloves in unison, then smartly slapped the Air Force F-15s across both cheeks. Then they about-faced and marched back to their cockpits.

Rivalry, one might say, is keen.

On my day to fly the F-4, I waddled out onto the ramp in the Louisiana summer sun, cooking in my Nomex flight suit, pressure pants, survival kit harness, helmet, and sword, and climbed the steel ladder up to the cockpit of the F-4. They call their F-4s "Rhinos," standing thick-skinned and threatening nose down, the horny surface near the cockpits worn bright by the footprints of many.

Crawford's crew chief, who was standing up on the wing with a proprietary air about himself said, "you can step anywhere." He took my hand, helped me down into the pit, which is an apt name for the deep, dark hole under the canopy and behind the pilot where the GIB (Guy In Back) sits.

We spent considerable time in the hot sun strapping the F-4 to me. Many straps, many buckles, no two that looked alike or functioned the same. And I hadn't just strolled out there. No, there had been an intense half-hour briefing from Lt. Mike Cantella, showing me the three different ways of emergency egress. Of them all, the most simple way is to just pull the trigger on yourself, but don't wave bye-bye to anyone if you ever intend to use that arm or shoulder joint again.

"Does heat, confined spaces, and bondage bother you, sir?" asked the chief over my cockpit.

"No," I answered honestly, "but you need to know I have a well-proven endurance of about 20 minutes in airbatics."

Pilot and crew chief exchanged grins like a pair of bayou alligators as the chief placed a pair of sick sacks under the top layer of the many belts over my belly. They both knew this was going to be about an hour-long mission to go out over the Gulf of Mexico, bounce some Air Force Eagles, and mix it in mock combat.

My great moment for looking like a fighter pilot was at hand. Next to us, Pilot Jim Mathews and his GIB Jeff Randolph were tucked in for a formation takeoff. And I do mean tucked in. You could have walked across, and they stayed that tight all the way out to the MOA airspace over the Gulf.

Before the ground roll began we looked at each other, visors raised. I gave them the big ol' "thumb up," just like in the movies. Oh, boy!

Down in my pit I felt what was to be the first of many and various gravity-busting pushes from the seat of the Phantom. As thrust shoved against my back I looked over at Mathews' F-4. When he lit his afterburner the whole tail-pipe section of his airplane turned bright cherry red.

I was glad they had dual controls down in the pit so I could follow through on the stick, rudders, and thrust levers. No mixture control. It comes, you might say, ready-mixed in the can.

"You got it," said Crawford a few minutes later as the Louisiana coastline slid away behind us.

"No, sir," I replied. "Too close." I had enough formation flying experience to know I didn't know anything about it. Mathews' plane was painted to us like a beautiful reflection. In the front cockpit Crawford nodded and did a lovely slow roll away from his wing man, putting space in the sky and the sky in my belly. Ulp!

At last I laid a reverent hand upon the stick of a real fighter plane. Firm. About the same pressures as a 182, only it can do wild things with sea, sky, and horizon with just a flick-over.

In any strange airplane I first try to fly straight and level. I went on instruments at once—everything is the same color up where Phantoms prowl. Little dollar-size instruments filled the panel in a random pattern. I got them all settled down except for one, which kept going round and round. I leaned forward to see what it was. It was the altimeter. The F-4 was loafing along upwards at about a

45-degree climb. I pushed over as easy as I could. Ulp!

Then the Air Force Eagles found us, and the fighter jargon began in my helmet. "Two, five o'clock high." We pulled into a steep evasive turn, the pressure pants lovingly squeezed my blood back up into my brain. Then, two F-15s started a pass at us from the other side. "Seven o'clock level! Oh beautiful! Bax, you see that?"

Yeah, just a glimpse of it. My hands were now locked into a cutout hole that somehow had been thoughtfully left in the top center of the instrument panel of the pit. Perfect for hanging on. My head, which weighed a ton during these maneuvers, had now come to rest down on my forearms. We chased Eagles. I closed my eyes, tasting salt water. Madness, all madness, upside down, trying to hold the ocean up inside me, listening to the happy terse radio comments between pilots. Fine fighter pilot I was. Sky full of planes, twisting, diving, climbing, and I never saw a one of them.

Mathews, alongside, to Crawford: "Bax has his head down."

Crawford to me: "Bax, you okay?"

Me to Crawford (needing to be honest but not wanting to spoil the party): "I'm hanging on."

Jet fighters, ours and theirs, kept slicing at each other in the soundless sky. The Gulf MOA looks big on a chart, but it shrinks to a pretty small sandbox for Eagles and Phantoms to play in.

I raised my head and got a glimpse of myself in the rear-canopy bow mirror. Maybe the sight of myself in helmet and goggles in the cockpit of a real fighter plane would at least give me heart.

Ghastly. In a G-turn, face long as Bob Wills' fiddle.

Crawford in my ears: "What do you want to do, Bax?"

Me: "Go home." I had lasted nearly an hour.

Home was a streak. Incoming fighters usually signature the runway, a little something for the ground crews.

Me to Crawford: "Just straight in. Please." I had the sick sack at the ready, eyes shut.

Crawford turned wide and gentle, touching all those Rhino tons down like a Cessna 152 on a cool day.

He was popping the drogue chute.

"We down yet?" I asked.

"Thanks, Bax," he grinned.

Our crew chief fished me, pale and soaked, up out of the pit. My only honor was to hand him the sick sacks, still sealed.

"What do you need, sir?"

"Just get me horizontal, but soon."

They laid me out in bliss on the couch in the squadron's ready room. Fighter planes zoomed at every angle from the paintings on the walls. I closed my eyes.

"The right stiff," I was thinking in self-contempt. "All chuck, no Yeager."

When I came-to a couple of hours later, the room was full of fighter pilots, their day done and a party going full blast. Squadron Commander Lt. Col. J.J. Case was sitting on the couch next to my head. Sitting guard over the body. I looked up and met his compassionate brown eyes. "Now you are one of us," he lied lovingly.

I arose, and silence fell in the room. With all that was left in me I began to sing, "Oh, there are no fighter pilots down in hell . . . ."

There were cheers. They roared the song along with me.

Fighter pilots are, indeed, a special breed of men.

"Cajun" is what the rest of the country calls the Louisiana French. "Coonass" is what they call themselves. Later, I got a wall-size photo of a Rhino over the crescent city of New Orleans. On the wide matting was an inscription declaring me an "Honorary Coonass Fighter Pilot." And everyone had signed their names to it, all the way around the photo. Which now hangs among my few other wall treasures.

27

Skydiving Pigs

LETTERS, as I have said before, are important to us at *FLYING*. Not just cast aside, but sent round the chairs for all the editors to read, copies sent out to any of us stringers who were mentioned. Mistakes we let slip through get the most mail and the most gleeful hootings.

But I enjoy my mail and read it for story ideas, too. Here is one that ain't quite gonna make a story. It was forwarded to me on a regular "claim and loss" form, with a memo that said, "Gordon, thought you might to want to check this one out." And signed. We throw out unsigned mail as junk.

The report on the claim-and loss form read, in part:

—left wing, seven holes, left aileron two holes . . . right wing four holes, right aileron one hole and buckled. Left engine baffle two holes, spinner one hole, big nick in the propeller. Fuselage 15 holes, mostly down each side, all windows broken out except left aft. Altimeter, tachometer, and airspeed indicator smashed, mixture control and throttle damaged, radios damaged. Vertical fin two holes, rudder one hole, horizontal stabilizer three holes, rotating beacon smashed.

Each reference to a "hole" indicates damage inflicted with what was apparently an ax.

Will be sold as is, where is, to the highest bidder. The company reserves the right to reject any or all bids.

I held this letter in hand for awhile, visualizing this chopped-up, single-engine, low-winged aircraft, and just as easily came up with a mental image of the ax wielder. Although I have a great and natural curiosity about all things in aviation, I found it easy to just lay this letter aside and forgo the idea of getting in conversation range (or ax range) of a person capable of such passion. This is another and more elaborate way of saying I chickened out. But haven't you ever felt like taking an ax to an airplane? Would be kind of grand for a while there, wouldn't it?

A briefer letter, and more on the happy side, came from Tim Carbury of Hampshire, England:

> I have been involved in the restoration of an old Tiger Moth. Well, the lady finally took to the air again last year, and I am currently trying to reach the stage of where I can fly her solo. I think I am getting close.

Then, along with his letter, Tim sent a 5″ × 7″ color photo of his lovely little stringbag, taken from a three-quarter rear view. The soft white of the Moth's wings and the dark green trim of the upper fuselage, flying above the misty English countryside, is so lovely that the photo is now window-mounted just above my typewriter. Not enough here for a full story, but a fine gift, a picture that calms me whenever I look at it. Thanks, Tim.

Then there was the terse postal card from Penelope K. Amabile, re my spelling in a Mooney Club public letter.

> Amelia Erhardt??? Perhaps you mean Airheart? Aehart? Airhard? Erhart? Try Earhart, oh you champion of women pilots.
>
> (signed "Affectionately" anyway.)

And the other day somebody asked me why I was scared of women.

Not all. There was the Maiden of Oshkosh who shyly stopped by to visit us in our press room one day. She told us of her hopes and dreams to someday to be able to afford her own airplane and learn more of flying. We kept in touch via mail. There was her letter of triumph; she had acquired a Decathlon, and matching letterhead on her stationery. Later there came a tear-stained letter telling of how a hailstorm had beaten her wings down to the bare spars. She was not at Oshkosh that year.

And the guy who wrote to tell us of what a faithful subscriber and reader he was and that we had sent him a copy of *FLYING* that was hopelessly scrambled and some of the stories missing. Now, how can I tell you that all of us in Editorial are a long, long ways from the plant that actually does the mechanics of printing. And that the subscription department is in far away Colorado and they know us not. No, all I did was send him my own copy of that month's magazine and tell him to keep the screwed up copy, it might be a collector's item someday.

And I keep up a delightful exchange with Karl-Eric Gutenwalk, Dino International, Sweden, who flies gliders in the Alps and keeps warning me of aircraft corrosion.

Some subjects never die out. They run on like a continued story, wanted or not. The business of skydiving animals had gone on steadily since 1980, with no encouragement from me. I can't find the clipping now about the Great Florida Pig-out. Sounds more like something that would come out of California. Jumpers leap carrying porkers in their arms. Reported to the FAA, the Fed is supposed to have said that he thought the matter was disgusting, but that there is no accounting for what some people will do.

Then came an absolutely fascinating letter from an ex-Marine/ex-Green Beret/Chief Warrant Officer (Ret.) who goes only by the name of Sneaky. We printed that one, signed or not, for the seldom-seen insight on our fighting men from Vietnam.

"Sneaky" tells me this 'Nam sea story of a pig being transported out of Da Nang and up Happy Valley in a big, green, rotary-wing aircraft. The porker, it seems, was a part of the load manifest, supervised by a crew chief and door gunner to be known here only as "Buck." The value of the story is the authenticity of the language. (That's my excuse.)

The pig failed to make roll-call as the chopper unloaded at the forward base. No reason was given. Pigs, whiskey, and other perishables came and went in the chancy business of air transport in that war.

However, upon returning to Da Nang through the usual ground fire, the whirly was met on the ground by many heavy-metal Air Force types who were all bent out of shape about a pig that had penetrated the wing of one of their parked VNAF DC-3s; it had dropped out of the sky from about 3,000 feet causing considerable damage to the Gooney Bird and strike damage to the pig.

Buck's side of the story was that he saw the pig working his legs down through the bottom of the wicker basket in which he was being transported. "The pig then shuffled to the door, looked out, looked back at Buck, winked at him, and jumped. Basket and all."

The story has since grown to where the beast was a cow, but Sneaky says it was a pig; he was there. And only Buck, wherever he is, knows what actually happened. And it is faithfully reported back that neither a bottle of Scotch nor any other mind-altering refreshments will cause Buck to change the story.

Sneaky says he always turns to my column first when he gets the magazine. I told him I do too. It's the only way I have of finding out if I still write for them.

28

Concorde

"Here he comes!"

Five hundred thousand heads turned and stared south into summer skies. The sharpest eyes saw a black speck, coming, coming.

There was some jostling, a pressing forward of the mass, as if a foot or two closer would be better. Long lenses were raised, a battery of more long barrels than ever defended London during the Blitz, all swung toward that fast growing speck in the sky.

"There he is!"

A flash in the heavens, the sun. Then I realized I was looking into the power of his landing lights. Now we could see his delta shape, a ghost of bright metal hanging in the sky. No airplane has ever landed before such a sea of people. At Le Bourget in 1927 Lindbergh wrote that he first thought the dark mass of night shadowed the airfield, then realized it was a moving, living crowd of humans. He said he was astounded. He had spent 33 hours crossing the Atlantic ocean alone.

The Concorde left London before lunch, deplaned passengers at New York, then came on into Oshkosh before dinner. "I'll see you about 5:30," said the jaunty Captain Cook of the Concorde.

We were a solid mass of people out on the ramp at Oshkosh. When we sighed, we sighed together. The 6,702-foot runway would be one of the shortest that this great airplane had attempted to land on. I was full of apprehension.

My mind raced, recording every detail to tell later. Of how the great puff of burned rubber curled up into his wingtip vortices as he touched down. My God, such an airplane. Our heads swung to his oncoming rush, pilots amongst us thinking, "He's too high, too fast. He won't make it." I worried in utter silence, it seemed, as the utter beauty of this long-stemmed white ballerina filled my eyes.

Then a sudden crash of sound! Women screamed, children fell to their knees, cameras dropped at the chest-thudding roar as Captain Cook put her into afterburner power. "Oh lordy, an abort. Please let him make it!"

And just as suddenly he filled our heads with the arrowed vision of thunder, and he was gone, arcing away in a gently climbing turn. If I could have heard the PA system speaker, which was only a few feet away, I would have heard Captain Cook say from his cockpit that this was only a touch-and-go. "Bumps and circuits" as the British say it, and in the Concorde, yet. Oh Cook, you master.

He made three more low passes, showing off each side of the Concorde to the huge enjoyment of himself and every shutter finger. I'll bet enough images burned on 35mm film to have reached across Lake Michigan.

I now knew the next approach would be the real finale. His flapless wings flared before our eyes. To me he looked a little lower, a tad slower. The puff of rubber smoke appeared to be on the numbers of Runway 36 this time. He sizzled by us at midfield and still had enough runway left for an intersection turnoff. As he turned the great white shape towards us, as if taking a bow, I wanted to laugh or cry or hug somebody, do something. We were swept up in mass hysterical joy. I would have led the rush to carry Captain John Cook away on our shoulders.

Oh hear this, ye cryers of doom, ye bean counters of the powerless descent of the fortunes of aviation in America. There were more of us there to see the Concorde arrival than Pickett had at Gettysburg. Just a simple thing, to see an airplane land (not even the biggest airplane) had gridlocked traffic over half the county. And the talk of it filled our mouths all that night.

We talked about aviation. We had been gathered at the grass roots of aviation, the EAA convention at Oshkosh. Here it comes, just a speck now, that only the sharpest eyes can see . . . .

29

Hello Stearman, Old Friend

GEORGE MITCHELL of M&M Air Service asked me to come over and see the new airplanes. He doesn't do that very often. George owns the Stearman two-holer I've been flying on-and-off for the past 23 years. I've flown it since it had the original 220-hp Lycoming it was born with, then through some awkward times with the 300 Lycoming they hung on it next. I say awkward, but there were only two conditions when that 300 Lycoming was hard to start. When it was cold. Or hot.

We always started that high-compression engine by hand. It would leave big, strong rice-fed men lying out in the grass, puffing for breath, waiting for the strength or the will to come back again.

Then George put a 450 Pratt & Whitney on it. That engine was for the airplane made. Surprisingly easy to start for something so big and broad-bladed, that low-compression engine would always come rumbling to life after a good quarter pull on the propeller. I had flown other 450 Stearmans, but not the M&M two-holer since they put that fine engine on it.

The reason for the long delay in the engine change was that George kept all of his 450s busy on his sprayers. There just never was one to spare for the old two-holer (which is all we ever call it). Out of a flock of Stearmans that once numbered up into the thirties, the Mitchells kept the two-holer as still-stock. It was their check-ride airplane and was also used for short ferry hops around the rice field strips. Now and then somebody just took it out to play. All the others soon had high-lift wings, which were not good for anything but duster turns or carrying massive loads in level flight. The old two-holer would still do all it used to do when it taught Army Air Corps cadets to fly. A noble airplane, much beloved by all of us.

George had been in a period of transition, first to some Ag Cats, and then the gosh-awful investment in the first of his turbine-powered airplanes.

There were several reasons. The big round engines were getting very scarce. And not lasting very long, even out of the best re-man shops. Some say it was just "tired iron," although I'm not sure there is such a thing. Whatever, he was making a transition into propjet-powered ag planes now and wanted me to come see.

He had one of the first Air Tractors with turbine power, and a huge turboprop Thrush. Designed for the big radials, each airplane looked like an anteater with the tiny turbine engine way out in its pointy nose. Fast, they were plenty fast. And the length of time between overhaul was now measured in thousands of hours instead of hundreds. And the turbos were quiet.

As population began to crowd in on the rice fields, Mitchell was always getting calls about noise. "A big Pratt & Whitney would dance a coffee cup across your kitchen table. Now, when these turbos go over, ZZzzzzzz, they just think it's their oven timer going off."

George was laughing, but he is serious about being a good neighbor. He is the second generation of Mitchells to run this operation and was a pioneer in the movement to organize the Texas Air Applicators and the National Agricultural Aviation Association. A past president of both, he has seen ag aviation stand together, not so easy for environmentalists and federal agencies to pick off now.

A pilot with about 8,000 hours of cockpit time and a B.A. degree in business, he and Gail have five children, the oldest now starting to fly. Now weathering the shrinkage of rice field acreage, George is, as always, trying to think his way through it.

George is also a sensitive man. He saw me looking off across the field to where a pair of his Stearmans were coming home to roost; I recognized one of them as the two-holer. Must have been some kind of a ferry hop. We grew silent, watching the Stearmans touch down. He knows how I feel about that airplane.

Finally, George broke the silence between us. "Why don't you go over there and fly it?"

There was a strong gusting wind, and as I drove around the field I had plenty of time to think how long it had been since I'd flown the two-holer. That sort of gave me the cold collywobbles, but what the heck, I've always gotten the cold collywobbles. The Stearman has a way of teaching you what you need to know very quickly.

There were two other Stearmans still working the narrow strip and coming up to the loader truck. Pilot John Burrell was about to park the two-holer by the hangar, recognized me, grinned from his cockpit, and beckoned me to come on. He left it idling as I climbed, against the prop blast, into the front cockpit. That was another spook. I don't think I'd ever flown one from the front cockpit.

The pair of working sprayers were landing into the wind, taking off downwind. That's not unusual in ag operations, but I hadn't done a downwind takeoff since Ike was President.

"Let's go." said Burrell.

"From here?" I asked.

He nodded.

OK, no time to find out like right now. I ruddered around, aimed downwind, far enough over to clear the loader. Both the ag planes were in the air by now. I opened the throttle slowly so the torque swerve wouldn't catch me off-guard.

The field was wet and soggy, with a row of telephone poles down the right side. All I can figure is the wind got down on top of my tail feathers. I staggered off, swerving toward the poles, too much correction for the expected torque. We were off, but drifting fast toward Ma Bell's woodwork. The tail wanted to stay down, and heah come de poles, heah come de poles!

I felt Burrell's sure hand and foot on the controls. We did a low swoop away from the wires, then he gave it back to me with an odd look in the mirror and a shake of his helmet. Don't ask me, Burrell. I never felt an airplane take off like that before.

We circled around to line up for the landing. It had been a long time, and I wanted to show Burrell a thing or two to make up for that takeoff.

To see past all that Pratt & Whitney I came in with a gentle slip. Actually, the Stearman lands better with the heavier 450 engine on it; there is no tendency to float. You get used to being blind.

On the flare I knew I had it made. A greaser. All three tires of that BUF touched and rolled at the same time. Looking up at me in the mirror as I laughed, Burrell shouted "What do you want to do now?"

"Same thing. Show you that one wasn't luck."

He nodded. I taxied on upwind a ways to give us lots of room for another freaky downwind takeoff, ruddered smartly around, and opened the throttle. Same damn thing. I felt strange forces pushing the plane and seemed to be paralyzed to stop it. Again I felt Burrell's hands and the controls and throttle wide-open. And again he prevented us from becoming bad news to AT&T linemen. I was sweating. The more I fly, the more I have to learn.

Topping the poles by a wide margin, Burrell took his hands off, and I danced us around sweetly to another short final. (Short final in a Stearman is so short you could hit the field with a baseball.)

Again, I felt I was where I ought to be: loving the song of the wires, feeling the measured decay of lift in her bright yellow wings, and lifting the nose at the proper instant—just enough to roll all three on the turf without a trace of lift left in the wings.

Burrell looked up at me in the mirror and nodded.

In the rain-soft turf the Stearman taxied like a big iron horse, 450 Pratt & Whitney giving an increased soft rumble to power us around the turn and back to the hangar. I was in no hurry to get out. I sat there and listened to her idle for a moment. An idling 450 Pratt says "chalupa-chalupa, chalupa-chalupa . . . ," the tall, handsome, old biplane rocking her wings to the nine fat pistons chasing themselves around up front. Then I cut her, listening to the last nicker-nacking of the pushrods and tappets, and the delicate tinks of the exhaust manifold cooling in the silence.

Burrell sat with his head down during all this, aware I was in my church. Then the two new propjets passed over for landing. ZZzzzzzz. They do sound just like oven timers.

30

The Past Fifty Years

AVIATION WAS SO TOTALLY EXCITING TO ME in 1933 that I could not imagine a person who was not interested in flying, and if there was such a person, then what was the matter with them?

I can't find any comparison to make today for how we felt about flying then. I was ten years old and every kid I knew wanted to grow up and be a pilot. Real airplanes were scarce and seldom seen in my hometown of Port Arthur on the Texas Gulf Coast, and real aviation magazines were even scarcer. But we would buy and read anything with an airplane or an aviator on the cover. What we got were the dime pulp stories of World War I. Rip-snorting reading were the pages of *Wings*, *War Birds*, *G-8 and his Battle Aces*, and our favorite because it was regular magazine size and had pictures, *Flying Aces*.

The discriminating critic could lay my monthly output in *FLYING* alongside the pulp novels of my youth and readily identify my school of journalism. The sun flashes on the pilot's goggles, the moan of the wind in the wires, tires kissing the grass, it's all there. I forgot very little of it and only left out the "slamming of stick and throttle hither and yon." I was a pilot many years before I became a writer and had long since learned not to slam anything in the cockpit. The old pulp writers such as Arch Whitehouse could be forgiven for slamming. I'm sure none of them ever flew a SPAD or Fokker except everlastingly on paper.

As children of the first generation of pilots, we found good enough photos in the old magazines and newspapers to draw up our own three-dimensional plans of the airplanes and make solid balsa wood scale models of them. This was years before plastic models, and although there were a few kits of "flying models," these stick-and tissue jobs were too painstaking to build and too frail to play with. We had aerodromes of cardboard-box hangers and flew dawn patrols. The sounds we could make to imitate a rotary engine or the stutter of Vickers machine guns were the continuing sounds of happy kids at play.

My dear old Mom encouraged this love of aviation. She went to a lumber store and bought a two-by-six piece of balsa wood from which came many a fine model plane in living colors. Mom took hard-earned grocery money for the single-edged razor blades, sandpaper, and paint, and gave up the use of a big old table upon which the phantom outlines of airplanes were to remain forever. My room was my own; she never tried to clean up around a model in construction. I was a lucky kid.

She gave me a dime a day for school lunch, and I ate chili because it only cost a nickel and the crackers and water were free. This saved me a nickel a day and by week's end I was fat with 25 cents cold cash. Real wealth in 1933, age ten. I began to save up to buy a "mini-camera," which cost $3.50, and with this began to photograph the models in real-life settings. I still have some of those fuzzy old black-and-white pictures—the squadron in front of the hangars, or fighters twisting and turning in deadly dogfights, hanging from invisible thread. For a shot-down fighter I would twist an old stocking of Mom's around the thread—looked just like smoke from a spin-and-burn plane. A fondly remembered time this, when my only interest in a silk stocking was to use it in table top photography. Life was not so complicated and no less rewarding.

I soon had a large collection of scale models, and Mom let me hang them from the ceiling of my room. Each model hung from a heavy white thread, attached to the airplane with a hook-shaped pin. The thread ran through a staple in the ceiling; then all the threads ran to one big staple over at the wall. At the wall, all the gathered threads ran down to a system that allowed me to raise or lower the models all at one time. By hooking the gathered threads on a nail the models would stay at their regular flying place just below the ceiling.

Mom was proud of this. She would bring company in to show off Gordon Jr.'s room. While the ladies were looking up at the models, Mom would give me the secret sign, and I would let them all down in one swoop. The visual effect was that the ceiling was coming down. Folks would duck and scream. Like I said, Mom was proud of me.

Once she hired George Moon, a young photographer in town, to come out and take pictures of the room, the airplanes, and me at the work table. Photographer Moon liked it too, confessing that he was taking flying lessons in a Piper Cub out at the newly operational Parker's Airport, a pasture on the outskirts of town. Moon took me flying in the J-3 as soon as he got his license, a day never to be forgotten. He let me hold the stick awhile and fly the airplane.

Today, Parker's Airport is a subdivision, but George Moon and I kept in touch from then on. He started up the aviation department of Gulf Sulphur with a Navion after the war, grew into a Lockheed Hudson bomber converted to one of the early business exec aircraft, and eventually was based a Westchester, White Plains, New York, flying the company's Falcon Jets.

Moon would still invite me to come fly with him whenever our pathways crossed and long after I too became a pilot and was writing for *FLYING*. Aviation friendships are often like that. Riding with him in that Cub—and in the Falcon about 35 years later—was a dream come true.

Jim Robertson and I also began a lifetime friendship back then at age ten. A neighborhood kid, it was hard to find another boy who understood "plike," or "play like." He had the imagination to build and fly the model airplanes. We called each other "Gates," still do, from the popular radio comedian Jerry Colonna's show opening of "Greetings Gates." It takes a new wife some time to realize that we are both Gates, although neither of us is Gates.

Gates and I "flew" our bicycles to school, coming into town on empty rural roads, weaving, circling, each trying to get into the deadly six o'clock position on the tail of the other. Banking, weaving, tat-tat-tatting our machine guns, being a deadly Fokker tripe or a steady S.E.5A. We punished our bikes, flew in our minds. The young Air Corps should have had some way of finding guys like us, just channeling us right on into becoming fighter pilots. We would have been American Kamakazis by graduation time in 1942. But the Air Corps was stooling around with something about so many years of college education. Wanting gentlemen and officers, I guess. Missed both of us. Pity. We were simply ready to fly and shoot.

Gates and I read the story of the Wright brothers at the same time, and were both inspired to build our own glider and fly right then and there. "Fishflakes" was the glider's name, because it sounded zany enough and was also the name printed on the wooden box we used for a fuselage. The glider had a 10-foot-span externally braced low wing with slight sweep back. The tail was carried on twin booms, that is to say, bean poles, extending back from each side of the fuselage. Control would be by wing warping and would be added later, after we found out if controls would be needed.

Mom gave us all the wax paper we needed to wax the shingles of the barn roof to get all the launch speed we could. She also showed us how to make flour-and-water paste, with which we attempted to wrap butcher paper around the airfoils and tail. Wing camber came from the placement and height of the wing spar, a practice still in good use today, but trying to get that stiff paper to stick to those rough-surfaced bean poles was maddening. At last, Fishflakes stood ready for flight.

At school we had done a lot of talking about the glider. Or at least *I* had. Jim is the silent type—a doer, not a talker. I had formed an aviation company and was taking orders for future production with plans to enter into the aircraft building business pending the success of the first flight. My hype-ahead-of-performance style can still be found in the aviation industry. I was making poor grades, failing arithmetic, but what, me worry? Soon I would be circling above the schoolhouse above all that, the idol of every girl.

I honestly can't remember which of us flew first. No matter. The flights were close together in time and identical in results.

Each of us wore our flight jackets, helmet, and goggles, which all the stores sold for boys then. The jackets were black, made out of the same stuff used for kitchen table covers.

I had been worrying if we had enough bracing in the wing struts, top of the box out to about one third of the spar, but the wings never showed any signs of stress. Fishflakes just rumbled down the wooden shingles, gaining speed, then tipped off the edge of the roof and went straight in.

"Whomp!"

"It didn't get hurt none. You try it."

Another rush of wind, the "almost" feel of flight, then "whomp."

So bitterly disappointed were Jim and I that we left Fishflakes were she lay at the base of the barn wall, in the weeds to decay while we designed and built a very successful little boat and took up sailing. Jim never lost his love of sailing and had a long succession of boats, each larger than the last. Now in his retirement, he could be modestly called a yachtsman. Jim's face never changed. Photos beside his boat would be recognized as the same honest-faced kid of ten. He sails Corpus Christi Bay and flies with me in my Mooney.

I pursued my love of aviation, although we knew little of the commercial aircraft of the day. Once, we discovered a beautiful new OX-5 Travel Air biplane in a rural hangar. Respectfully, and in awed silence, we examined the beautiful ship without touching it. We decided it must have been some kind of Fokker, although we had never seen a picture of a Fokker with two cockpits and such a beautifully cowled engine. We simply didn't have anything about civil aviation of the day to read and learn from.

But when Clarence Chamberlain barnstormed through in his Curtiss Condor, I knew the Air Corps had a bomber of the same design. A box tail and huge V-type liquid-cooled engines out on each lower wing. Mom gave me five dollars for the ride, although five dollars in 1933 would buy a week's worth of groceries at the company store. I had a seat right by the engine, and I'll hear that Curtiss D-12 for so long as I live.

And I admired the way the cabin roof of the Condor curved down steeply to such a narrow windshield and a little cat's nose out in front of that. The Condor's face looked just like the face of my old black tomcat—strictly business.

Flying bloomed for me right after the war. I came home from overseas, had a family, a job, and still the burning desire to become the image of my own hero, a pilot. There are some twenty-minute flights in my first logbook, all I could afford on those days. And my first passenger after I got my license was Mom. Don't you know she loved it!

But something had changed. In the postwar world, pilots were just ordinary people, some of them even women. But to be at the controls, high and free, was all I had hoped it would be. Still is.

FLYING Magazine entered into my life in a very vital manner in those early postwar years. I had been licensed without any ground school. The emphasis then was mainly on stick and rudder. Most of us had no radios and navigated by map, compass, and watch. (By railroads, too, to keep this honest.) But as I continued to fly, airplanes and airports became more complex, and the "How To" stories in *FLYING* gave me the basics I could have never found otherwise. I came to trust the magazine, to learn from it. Another dream come true was to start writing for *FLYING* in June 1970. I never became one of the experts or product testers. I seemed to simply fall into the role of story teller for the magazine, the same thing I do today.

For the purpose of telling this story I got out a June 1970 magazine and laid it side-by-side with a June 1987 copy. The more things change, the more they are the same. The most noticeable difference is today's bursts of color compared to black-and-white photos and pages of good grey print.

The table of contents reveals some evergreens. "Flying Mail," "Editorial," "On Top," and "I Learned About Flying From That" have lasted all the years, along with the continuing updates on new products and basic techniques. Lots of names have lasted too. Senior editor Dick Collins is now Editor In Chief; Peter Garrison and I are still doing what we always did. Other names from early books read like a Who's Who of aviation journalism. Dick Bach and Ernie Gann wrote short bits for us then, and the list of staffers from 1970 who have gone on to riches and fame elsewhere shows *FLYING* to be the cradle of achievers in aviation writing.

An overview of the history of flight in my lifetime has shown that the only constants are surprise and change. The DC-3 and the J-3 Cub were revolutionary surprises back in the depression years of the thirties when so many of us then, as now, could only dream of flying.

Today, with millions coming to the airshows just to look, I think the stage is set for another round of surprises. Man's desire to fly has never been long denied.

31

The Great Mooney-vs.-Boeing 727 Race

BACK IN THE BALMY SPRINGTIME OF 1972 when the fuel crisis was at our throats, I raced *FLYING* editor Dick Collins in a car-versus-airplane contest. We called it the Great Fuel Race: Grumman Tr-2 versus Ford Pinto, these two chosen for being the most miserly of vehicles.

We flew a triangular course of 289 air miles, from Shreveport to Ruston in Louisiana, then to El Dorado, Arkansas, then back to Shreveport.

In order to make the contest more real life, I insisted that Collins had to dismount, find his way into each town, and get a stamped time-and-date receipt from the local post office. Anybody can look good just flying airport to airport. I wanted the pilot to face the getting-into-town hassles.

The results of this contest were a surprise to us all. The issue was fuel economy, but Collins' Grumman used only one quart less fuel than my Pinto. The real lesson became apparent when Collins got back to Shreveport and waited for me. And waited. The sun went down; it got dark; yet they waited. Collins arrived two and a half hours ahead of the car.

The real message in this contest turned out to be not so much a lesson in fuel consumption (we both did good on that) but rather, "are you willing to spend five and a half hours in a car to accomplish what you can do in two hours and 57 minutes of lovely flight?''

At the time of the Grumman-Pinto contest, President Nixon had just asked a trusting nation to please drive 55 to save precious fuel. Out on the highway after a few hours of 55 mph I got utterly fed up with such patriotism and began to flog that little Pinto bean box along at a bobbing 70. I thought I would never get that thing back to Shreveport. They cheered when I staggered in at last, like I was some long-lost brother.

Following the highways gave me only nine percent more distance in miles than Collins in his point-to-point flight, and I could drive right up to the post offices, hassle free. But none of that helped. The real discovery that day was the unexpected vast time difference between a trip at 55 mph and the same trip at 110 mph.

The story went to print and the fuel scarcity never developed as badly as forecast. But the sudden and steep climb in the cost of fuel is still with us. The operating costs of all types of aircraft began to go out of sight, and suddenly, used and neglected little Mooneys became very valuable. All that was back in 1974.

This year, 1987, editor Collins called with another great idea for a comparison race. This time he wanted to prove that a general aviation lightplane, flying along at a steady point-to-point 150 mph, was faster than a jet transport that zips along at over 320 mph but has to stop and set you down to wait for a connection now and then.

In this contest, guess who got to fly the sweet brand-new Mooney 252 Turbo? And guess who got to fight his way through airline terminals and ticket agents? It was the Grumman-Pinto thing all over again. I guarantee you that if I ever get to be editor of anything, I will treat the hired help exactly the same way I was treated.

To simulate a realistic business trip we agreed on the national average of about 300 miles, with at least one airline-to-commuter connection, the final destination being a midsize city. We would meet in the coffee shop.

Collins knew where we were going, but he kept clammed up about it. Wouldn't tell me the name of our city until the last day, which is not unusual for business trips. We would pick up the new Mooney at the factory in Kerrville, Texas, and start the race the next morning from the nearby jet hub of San Antonio. Collins said I could use my green card credit any way I wanted to. I should have chartered a King Air right there in San Antonio and beat the pants off him. My journey began the day before the race, a matter of getting from home plate, Beaumont, Texas, over to jetport Houston, a Continental jet from Houston to San Antonio, then being met by a factory pilot in San Antonio and whisked out to Kerrville.

At the start of this trip from Beaumont I met an old friend and ticket agent who, upon learning the nature of this trip, narrowed his eyes and offered to get on the phone, call other agents, and set me up with some red-hot illegal connections on race day.

A legal connection averages about a half hour to hour layover to allow the passenger to walk from one end of a big jet terminal to the other.

Now you see why the canny Collins was not telling me where we were going tomorrow. No way could I pre-pave the route with the help of my friends.

Ed Trahan, the factory pilot, met me in San Antonio with a brand-new 252-TC demo ship to whisk me out to Kerrville in minutes. He asked me if I wanted to fly it but I never fly an airplane I don't know anything about unless we are in an instructor-student mode. On the trip, Trahan, who flies like he was born at it, kept asking about my old '68 Mooney Ranger. I later learned he was restoring a look-alike '66 model as a work of love and was gathering bits of information about the early ships with manual flaps and gear. He got good news from me about all that.

It took me all day, changing planes at Houston and San Antonio, to make the trip out to Kerrville, a run that takes me only two hours in our old Mooney. I was ready to go on home and write the story.

Flying his Cessna TC210, editor Collins breezed in at the end of the afternoon, looking fresh and frisky, although he had just flown solo in that Centurian from his home in New Jersey, through a weather system, and out to meet us in west Texas. At dinner that night with Mooney honchos Paul Kulley and Brant Dahlfore in a lovely setting overlooking the river, he advised me that our destination city would be Texarkana, Arkansas.

Well, not too bad, not too good. I was anxious to get back to my room, get on the ear pistol, and play the 1-800 game with the airlines in San Antonio to see what I could put together. We had agreed on an 8:30 A.M. departure from SAT, again a typical scenario for the airline traveler.

There is absolutely nothing that will wear the thin veneer of civilization off of me faster than a good session with 1-800 phones. There were seven major airlines listed in the yellow pages; I called them all, starting from the top. Lots of times I got the "agents are busy now, please stay on the line" recordings. At Continental the phone rang and rang, never answered.

It was only a little after 9:00 P.M. I guess they had all gone to bed. Wish I had, instead of leaning across it taking notes from the ones who answered.

American was nice; they had one flight too early and one too late but seemed to care. I had to wait out the recording at Delta, then a level-headed woman came on the line. They had a flight at 8:25 A.M. for DFW, with a Delta connection on into Texarkana by 11:05 A.M. There went the race, and I knew it.

Eastern was kind enough to tell me that only American and Delta connected to Texarkana and then asked where it was and would I describe it. I did.

US Air said the same thing as Eastern. I even tried Southwest, knowing they go into Dallas Love instead of DFW. "We don't fly to Texarkana at all, sir." Pity.

By this time I was phone-whipped. I took the Delta connection. That was one of the worst parts of the airline trip, and I could have avoided it if only one good travel agent was still open that late.

Next morning Collins and I saddled up the new 252 Turbo. He did not offer to let me fly it, nor would I if he had. We went from Kerrville to San Antonio in 15 minutes.

At San Antonio I said I had a close connection with Delta and was whisked over there by a little darlin' driving a company van. Delta never fumbled in getting me ticketed and off on-time. Somewhere out there on the airport, Collins was taking off in the Mooney. Wish we could have sort of clipped him or something, blown him away with jet blast.

The Delta crew gave me their numbers when I told them I needed them for the story. Cruise: 338 knots; distance to DFW: 225 NM. After we landed, their captain gave me his flight plan. I couldn't read it all but still have it. Nice chaps, all of them.

The forty-five minute layover in the DFW terminal flew by quickly, and somewhere out there in that Mooney so did Collins.

Again, a friendly bunch crewing the Atlantic S.E. Airlines commuter. They announced me as a passenger on the PA, and some folks came up and visited—a warm and friendly experience that helped to dull the knowledge that in the Texarkana coffee shop Collins was already sitting, waiting, the race over and won.

Collins held the turbocharged Mooney up to 11,000 feet, crossed the diagonal of Texas, and arrived at Texarkana almost two hours ahead of me. The Mooney beat the airliner by one hour and 53 minutes. He looked smug.

Collins offered to take me back home to Beaumont. The next airliner leaving Texarkana was at 2:15, would leave DFW at 4:02, and have me home by 5:12. Collins had me home at 1:15 P.M., time to sneak a good nap before anybody knew I was back.

Jet transport is unbeatable for long-range travel, but the personal business aircraft is unquestionably quicker unless the jet is flying directly to your destination.

The American business community did not delay in adopting the computer with its outstanding advantages in accounting and communications. We wonder now how long it will be before midmarket businesses recognize the small general aviation aircraft, either personally or professionally flown, as the answer to today's problems of time and distance.

32

But What Shall We Do With Ellington?

ELLINGTON AIR BASE was built by the Army Signal Corps as a World War training field in 1917. It was located on the flat, featureless, Gulf Coast prairie about 16 miles southeast of downtown Houston. The entire construction of this encampment, hangars, water tank, and all, took about 60 days. The book *Ellington 1918* notes, "Labor disputes twice disturbed the rapid erection of Ellington, but two companies of infantry dispatched from nearby Camp Logan brought a quick and amicable adjustment." Don't you just know they did. The bayonet blade on the M-1903 Springfield rifle was almost two feet long.

The first Curtiss aeroplanes were shipped in, uncrated, and flown on December 5th. A formation of ten flew over Houston: ". . . city streets were thronged. Every neck was craned skyward as the planes sped (at 90 mph) over the city."

The love affair between the southern belles of Houston and the dashing young cadets of Ellington had begun, and Houston kept a warm place in her heart for Ellington evermore.

There were only two types of aircraft at Ellington, the Curtiss trainer, not yet called "Jenny," and the first DeHavilland aircraft seen in the south. These were the DH-4s used in bomber training.

Ellington served a grateful nation through two major wars, the city of Houston grew out and sprawled around the flying field, Hobby airport was built just five miles west of Ellington, cadets became pilots, wars were won, and Ellington and Houston co-existed in an easy mutual admiration, demanding little of each other.

Ellington became a base for delta-winged interceptors, Hobby became an airline hub, and the two big airports lying so near each other became a problem to the growing postwar general aviation traffic. Westbound arrivals into Hobby would be just about on base leg when they had to overfly the approach end of Ellington's same-number runway. We Cessna pilots kept our necks on a constant swivel expecting to see some F-102 on final to Ellington drop out of the constant low-lying crud.

Nose up, gear down, probably with his head stuck in his radar scope, that hot shot could never see us. We never had a million-dollar crash at this crossroads, but we local pilots would pass the Ellington tower frequency around among ourselves and write it on the margin of the Hobby approach plate. This military tower was good about answering a quavering voice call from a Cessna and saying whether or not it was safe to cross the centerline.

The other problem is still common wherever big military and civilian fields lie close to each other. Some jasper would come struggling down the approach, break out, and find himself on final to 17 at Ellington. Hobby tower would go hu-hum and vector the errant VOR approach ace back over to 17 at Hobby. It happened all the time. Nobody fussed.

Then Ellington was closed during the Kennedy-era closing of military bases. Bulldozers, time, and weather set about leveling the familiar World War II barracks, but the steel-framed hangars, the broad ramps, and beautiful broad 9,000-foot runways stayed intact. A few of the old wooden buildings still stood in the empty checkerboard of cleared-off lots—the parachute loft and the old post theater still standing like distant white tombstones.

The ramp stayed busy, the hangars occupied, as outposts for a mix of Gulf Coast military. The Coast Guard claimed a part of it; so did nearby NASA. And various Air Guard units were based there or visited. The arrestor gear for tail-hook aircraft was kept operational with thanks from various Marine and Navy carrier groups operating in the Gulf. But Ellington loomed bigger and bigger as an underused piece of valuable real estate. What to do with Ellington kept coming up as an Air Force and municipal question. The Air Force solved it neatly; they gave the place to the city of Houston.

The timing of this event was ironic. Houston's large and active aviation committee had just been in the midst of studies about the need for a general aviation reliever airport. Traffic was, and always had been, heavy at handy Hobby. The huge new jet base north of town, Houston Intercontinental, had been in a construction and expansion mode since it opened. The city really needed another large reliever airport the size of Ellington. The only problem was that Ellington was on the wrong side of town.

Ellington, southeast of Houston, was surrounded by eleven airstrips, mostly privately owned, within a ten mile radius. They wished Ellington was northeast of Houston, but nobody has figured out how to slide an airport over, not even Texans. And so the problem remained, what to do with Ellington, only now it was a city problem.

Houston is the city of free enterprise, and free enterprise would get the problem. The city held a drawing to see who would get the contract to operate Ellington. Jim Bath, local pilot and internationally famed broker in turbine-powered aircraft, drew the luck bean. Bath is sort of the Lloyds of aircraft brokers. Arabian princes and Fortune's 500 beat a path to Bath. Jim is also sometimes a visionary. He saw great possibilities for Ellington.

In Jim's own words, "I sent out a letter to every corporation in America that owned an airplane, telling them of my newly opened Southwest Services, fuel, maintenance, hangars, ramp space, at uncrowded Ellington. You know what the results were? Zilch." Bath made a quick slicing motion with the flat of his hand.

The question now facing Bath was not a new one: what to do with Ellington.

Jim conferred with Ben Benham, his newly hired FBO manager at Southwest Services and also an ex-fighter jock like Bath. Retired Colonel Benham flew 227 combat missions in the Douglas A-1 "SPAD" over 'Nam, holds two Legion of Merits, 17 Air Medals, two Distinguished Flying Crosses, and taught Bath the tricks of the trade as a fighter pilot. At 6'4" the towering Benham is easy to find out on the ramp. As member of the same exclusive club, ex-fighter pilots, Ben and Bath put their thinking onto the same track, went for, and got a contract for refueling United States Government aircraft at Ellington.

Reports Bath, "There's not an airplane in the United States

inventory we can't turn around here. Business has doubled since we began this in 1985." Bath gestured proudly to a ramp, busy with F-4 Phantoms, F/A-18 Hornets, F-15 Eagles, and a steady parade of T-38 Talons from nearby Randolph Field, all in various stages of cooling and fueling.

But the sight that jars the mind is to see just as much traffic of little Cessnas, Pipers, Mooneys, and Beechcraft getting ramp service. The ramp has painted areas to sort of separate the chickens from the hawks. In an open manner I have seen nowhere else pilots wearing K-Mart mingle and marvel with ones wearing Nomex.

Not only does Ellington mother this mixed brood on ramp and hangar, but where else would a private pilot in a Cherokee 140 be advised there is an F-15 right behind him. The tower handles the mix smoothly, á la Oshkosh style. The only thing to set Ellington apart from any other big civilian airport is a notice to land long in order to avoid the arresting gear nested at ends of the runways.

Most beautiful of all this is the return of the airport kid to the fence. The big military-style main base gate stays open; the guard shack is empty. So when south Houston kids hear a fighter squadron arrive every week or so, they line the close-in fence, right against the ramp, eyes and ears popping, cameras clicking. A free society at work, at its very best.

Bath and Benham provide what they call VIP service for various Air Guard squadrons who blast in for exercises (dogfighting) in the radar ranges out over the Gulf. Jim provides them a regular squadron debriefing room for pre- and post-mission work. He urged me to come in and listen to a debriefing, to listen to the language of fighter pilots, still hot from the cockpit. "It's like nothing you ever heard before." He led me into the room with a just-returned group of pilots, his presence being my pass.

Eyes darted about the room, about the old civilian with notebook in hand. Never have I been in a room of such men. Uniformly small in stature and handsome of feature, their faces and bodies still tense from the mission just flown. This group had just fought each other with camera guns; the film briefing later would be classified, without old civilian reporters at hand.

Let there be no doubt in your mind that these are America's finest. No ordinary farm boy, no matter how graceful in muscle, can master the high-tech complexities of war on a multi-million-dollar platform traveling beyond the sound barrier. These men are the winnowings of the choices; far greater numbers are gently set aside for other less exacting work. I tried to envision them 25 years later, those who survive, plump and pink bankers, for gravity and time act the same on all men. I cast that aside; let me enjoy them as now, these are the point men. Some have recently seen enemy action; some soon will, in those little spot confrontations that make the news for a week or so. I would seriously pity whatever foreign fighter pilot gets between the blades of these scissors. In a language so tight, so special as to be mostly beyond our understanding, they go about their swift, deadly, and yes, satisfying business.

The debriefing room was small, intimate. Deep lounge chairs for some. Others had to stand up and pace. They were still wet with it, the imprints of oxygen masks still around their mouths, as they replayed the events of the last few minutes of combat. They seemed to have photographic recall of every moment of Hornets vs. Eagles. Not a drop of fuel was wasted here, dear friends.

Some stood at the blackboard at the front of the room and diagrammed each millisecond of the fight. Some just carved the air with their hands and words. Their language was so tight, so laconic, that I caught only brief bursts of it. Sometimes they would burst into brittle laughter; sometimes there were quick confrontations, deadly serious:

"First one we did was a ten. I think I was undectected, killed the left-hand man. Would have killed the right, but he broke away."

"I was looking for separation in the vertical flight. I pulled the trigger . . ." (Stands with hand on stick trigger, imitates sound of a 20mm revolving cannon.)

"We started at the same altitude, 30,000 feet, then you started that yo-yoing and were on your own."

"I didn't call a kill until my missles stabilized out. You called knock it off before I could call a kill."

"You went vertical. I got my speed up to stay with you. I couldn't even get a snap gun on you."

"There is no excuse to not be up there in that five-to-nine block." All heads turn to stare at the pilot, who seems to have lacked something. An instant, then a soft voice said knock *that* off, it could have been any of us. They stared at pilot, who took a deep breath.

"I couldn't leave from you, you couldn't leave from me with

the tech I had on board. It would have dwindled down to a salute and meeting another day."

"If a Hornet gets behind you, he'll probably stay there."

"I went to idle board, just squashed the airplane."

"I had the guns on you, one potato, two, I got you."

"I went straight down, right through Mach, went through boom boom."

"Did I get out?"

"You were out of it."

"Why did everybody stay down there at medium altitude?"

"It's that white Eagle. It stands out against the water."

"About that time I hit bingo." (Low fuel, came home.)

It was out of them now. They were standing, laughing, blushing, getting up. It was over. I asked one, "Do you think you have air superiority over any fighter in the world?"

Motion stopped. The room got deadly quiet. "Yes. But any fighter pilot would answer yes. The answer varies from day to day, would depend on what technology we were carrying."

I asked them about so many missed approaches on return. With great crowds at the fences the fighters would be on short final, then suddenly clean up, accelerate, climb, then bank steeply, showing themselves to our delight. "Oh. Oh, that. That's a precautionary maneuver. To see if the fighter still fits on the runway here after we get them heated up."

Like schoolboys they looked to see if I was writing that down, then tiptoed out of the room.

Ellington lives.

33

That Big Purple Airplane-Eating Cloud

IT WAS A WHOLE 'NUTHER WORLD, being a student back in 1956, and some of it liked to have killed me. I had never flown a nosewheel airplane yet, nor one with a radio in it. You'd be surprised at how much of today's FAA exams are about radios and radio navigation. My exam had none of that. It was all about flying and understanding the light-gun signals from the tower if you got into some airport that had a tower.

I think nosewheels were a good idea and brought in a lot of new students, but I sometimes think it was a dreadful day when we so let radios intrude into the peace of our cockpits and put some distant non-pilot fellow on the ground into the cockpit. Watching every turn, every change of altitude. Telling us which way to go and when we can do it.

I look in the old logbook, and there was hour after hour of practice in stalls and spins and 720 turns and chandelles and S-turns and slow flying. We were expected to know a lot about how to fly an airplane well.

Reading light signals from towers was easy, red for stop, green for go, and a lot of variations in between, including white. We acknowledged by doing what the light signal requested, or if in flight, gave a friendly waggle of the wings. That always seemed like such a gallant and aviator-like gesture, waggling your wings at the tower. In some places you could get arrested for doing that today.

We were expected to exhibit proficiency in all forms of navigation by dead reckoning, that is to say we found our way cross-country with a map, compass, and watch. This was actually sort of fun, a giant board game called You Bet Your Arse. In time we came to recognize vast parts of the United States from staring at them so hard. We memorized a lot of check points, and flying by recognition and memory of the terrain had the legitimate name of "pilotage." And, of course, we followed a lot of railroads and highways, but this was treacherous business. A railroad or highway could shunt you off onto the wrong road and if you were not paying attention to the compass too, could lead you far astray.

As a student I wasn't flying anything of over 95 horsepower, so those eagerly looked-for check points came along in a leisurely manner. There was lots of time to look out the windows and no reason to be flying very high. It was nice, except now and then in a headwind I could see cars on the highway passing me.

Staying low like that slowly unrolled the beauties of the countryside. We could be in touch with an occasional farmer who would wave to us and get a friendly wing waggle in return. Airplanes and aviators were admired by the general public back then. We had not yet become a social problem.

I worked out a variation on common wing waggling. I wagged the tail. This could be done by vigorous treading on the rudder pedals. As the tail yawed it would sometimes cause the propeller to cavitate and make real airplane sounds out of the steady putt-putting of those faithful little four-cylinder engines. We seldom had engine troubles back then. Crankcase cracking and fuel-induced valve swallowing were to come with the greatly improved engines of years ahead. Yet we were always looking for "a place to set her down." This too was a strict part of our training, a remnant of the not too recent past when most ships had OX-5 engines and none of these ran very long without some kind of a thrill.

My old logbooks show nearly all of my early work in aircraft of 65 horsepower. They were logged in as Piper J-3, Luscombe, and Aeronca. When we suddenly got the new-model Aeronca with the 95-hp Continental, we began to log that one in as the Champion. There was a serious difference in the way a Champion took off as compared to just a household Aeronca.

We were expected to know and be able to demonstrate four

different kinds of takeoff: soft field, short field, obstructed field, and plain vanilla. The 95-hp Champion really made a difference in all of these. You could briskly haul one up out of the weeds or mud—such flying fields were a part of our everyday lives. And I was also taught to do a sort of L-shaped takeoff, which was not in the books but was a sort of gift from my instructor. The L takeoff was for getting out of *really* short fields.

On the L takeoff I would lurk up against the fence close as possible, nose pointing 90 degrees from the intended direction of takeoff. Open the throttle against tight brakes, get her rolling as fast as you thought you could without tipping her over in the ground turn, then make the turn, come blasting out into the wind with all that draggy early rolling already done, lift the tail, and leave like a P-40. You don't see that sort of stuff anymore. I guess paved runways ended the need for it, and anyway it sure sounds like a good way to tip over a tricycle.

We were taught to do three-point stall landings, to lift the nose just as the wing stalled and touch down all three wheels lightly with minimum forward speed. This took some learning but gave a huge amount of satisfaction when properly performed.

The other kind of landing was called a wheel landing, or a "wheelie." This took a fine touch and a lot more room. Just a matter of flying the airplane down lower and slower in a level attitude until the wheels began to roll on the grass. Then closing the throttle and, with the stick, gently easing the tail end down. Such landings were good for strong-wind situations where you would not want to exhibit a lot of flared-up wing with any lift left in it. A wheelie was a sort of cheater's landing, but still a good thing to know how to do.

We were taught crosswind landings in real crosswinds, sometimes until the controls had come up against the stops and the student could learn about the limits of the airplane. "*You* be in control! Don't let this airplane fly you!" my instructor would snarl, his hand just inches off the stick. I learned a lot about limits then. The airplane's and mine.

Those early-day instructors taught by doing. Too much racket in the cockpit for much talk. We mingled a lot of juices. "Can an airplane be stalled at top speed?" asked one.

"No, sir, I don't think so." I knew that, if the question was being asked, they probably could be.

We were in the Luscombe, way up at 4,000 feet for this one. "Fly it as fast as you can on the level, then start a turn."

I did.

"Now make this turn steeper, steeper, tight as you can, and hold that nose up."

The Luscombe struggled to fly with all that bank and top rudder. From latter-day telling of the tale I think what the airplane did next is called a "vertical reversement." It snapped right over on its back, engine roaring, and then in a tight turn, but going around the other way. It sure looked like a good way for a SPAD to get a Fokker off its tail.

I never did it again. Nor forgot the message of holding top rudder in tight turns at any speed; 4,000 feet was a good place to find out about all this. Some learn it in the traffic pattern and make an awful mess.

But for all this heroic stick-and-rudder work, I still didn't know diddly-squat about weather flying. All I was told, and I was told this often enough, was to keep clear of clouds. Just like it says in the rule for VFR flight today.

The airplanes I flew had no weather-flying instruments of any kind. In later years I got my hands on a Cessna 150 that had an artificial horizon, but I remained innocent of its use. It simply was not taught to private pilots then. "Keep out of the clouds" was the main message. No instructor ever failed to tell me that.

But, you know, I couldn't see why a thing as light and fluffy and pretty as a cloud could be dangerous. Nobody ever explained "spatial disorientation," although I had read stories of pilots caught in clouds and spinning out. I just wasn't sure why. I knew from reading tales of derring-do about the early-day airmail pilots that some of them, above cloud layers at night, would see the city glow below and deliberately spin those old DH-4s through the cloud layer. Biplanes lose altitude slowly in a spin, and by spinning it on purpose they avoided an accidental spin and knew which way the airplane was spinning.

Worst of all, the only information I could pick up about thunderstorms was through hangar talk, or sometimes finding a story in *FLYING* about how dangerous those thunderbumpers are. But none of this abated my curiosity. How come clouds were so dangerous?

Sometimes I would deliberately fly through some of those little summer puff-ball clouds. "Whoof—bump." What's so dangerous about that? Clouds are beautiful. I would chase in and out of the cloud castles on fine summer days, but I did keep out of the big ones. I just flirted around in those majestic canyons. Oh, how beautiful.

It happened on my first student solo cross-country: Orange to Houston to Galveston and back to Orange. My old logbook shows a total of two hours and forty-five minutes; it was September 20th, 1956. I was in the Champion. What one might call a fat cat.

There was no Houston Intercontinental then, no jets either, no TCA—just lots of open sky. I entered the pattern at Houston Hobby, keeping a swiveled head and big eyes for the occasional Connie or DC-6. Tower gave me a green light for landing.

After Houston, down to Galveston, a great big deserted airport but right on the coastline with a sea breeze about equal to my touchdown velocity. I did a hover touchdown, landing laughing. From Galveston was the long leg back home. This would be my longest time aloft alone. I was starting to believe I belonged up there.

From a distance I could see the big fat old thunderstorm. I sat astride my pencil line across East Bay. Blue, it had turned green in the center, like an ugly bruise in the sky. The storm had a flat bottom, riding about 800 feet above the marshes. Beneath that I could see forever. A typical Gulf Coast afternoon thunderstorm. I decided that, instead of flying around it, I would just cross below it, right up close to its big shaggy bottom.

It was cool under there; the air smelled funny. Then, suddenly, I noticed the Champ was climbing. I thought that was odd and nosed her down some. Still climbing. A twinge of fear ran through me. You don't reckon a big old cloud like that could suck a little airplane right up inside it do you?

I shoved the stick forward and opened the throttle. Pointed down, engine roaring, I seemed to be stuck there, cloud tendrils reaching down for me.

Jeez! I dumped it. Rolled the Champ up onto knife edge where I knew it couldn't fly, and slowly began to fall away from that sticky monster. I let her fall, then began a steep side-slip, then leveled out just as the first hard downdraft shoved me toward the marsh grasses below. Now what? Nose up, throttle still wide open, darting like a minnow, I escaped.

Back in sunshine again I looked back at that monster cloud. Never forgot it. Oh, it was wonderful to be a student 30 years ago, just stick and rudder, and what didn't kill you made a better pilot of you. Believe me, we do it better today.

34

Super Connie

WHEN J.K. WEST CALLED and said, "Come out to the airport if you want to see a big old airplane," I went soon as I could. This is the same J.K. West of Angleton, Texas, who once let me fly his B-25 bomber, and the same J.K. West we had on the cover of our November '85 issue, flying that Martin B-26 bomber that belongs to the Confederate Air Force.

I was searching the skyline when I turned off at the airport, looking for "a big old airplane."

When the view opened between the two back hangars I saw it, sitting alone on the far ramp, it was a Connie, low, curvacious, and beautiful. J.K. wasn't kidding, the Lockheed Constellation is a really big old airplane. They are nearly all gone now. Replaced by the Lockheed Electra propjet, then the Lockheed 1011 wide-body, this was one of the first modern airliners and last of the piston-engine airliners. It had three rudders and four of the biggest radial engines Wright ever made. It was pressurized and had a long, round fuselage, curved as gracefully as a lady's leg. Lockheed first made them for TWA back when Howard Hughes had a lot to say about TWA.

Then known to the trade as the Lockheed 1049, she was a 1940 design, first made to compete with the Douglas DC-4. To demonstrate her speed and range, her maiden flight was a little cross-country run back in January of 1943. They made it nonstop from coast to coast in about seven hours. No other airliner could do that then. But the general public was not to enjoy this super airliner of the war years. All production went military, to the Army as the C-69 cargo/troop carrier. Navy buys came later.

During the war, still "keeping up with the Douglases" (their competition was now the Douglas DC-6), Lockheed pulled out their basic design and added a sixteen-foot plug into the fuselage, along with greater wingspan and bigger engines. Pratt & Whitney and Wright kept on developing bigger, more powerful, radial engines to keep pace with the growth of the four-engined airliners. Or it may have been the other way around.

Douglas entered the postwar airline period with their DC-7. Lockheed, now building what they called the L-649, named their new passenger airliner the "Constellation." Pilots called it "Connie," with much affection. Still do.

There are two ways to get more power out of an engine, other than by turbocharging. One is to get more cubic-inch displacement by adding more cylinders, which is what Pratt & Whitney chose to do. The other is to increase the bore and stroke of the cylinders, which was the Wright route.

Pratt & Whitney built a vastly complicated engine, sometimes called the "corncob" because that's what it looked like. The engine had four rows of seven cylinders. Wright opted for keeping twin rows of nine each, but, with huge pistons of about six-inch bore, got almost as much power as the big P&W.

The Wright saw wartime use in the B-29 bombers, but was plagued by engine fires. Called the R3350, the engine was viewed with strong mixed emotions by crews crossing vast distances of the Pacific with them.

By 1950, Wright had revised the R3350 into a turbocompound engine, using exhaust gases to feed power back into the engine through a complex system. Takeoff horsepower rose from 2,800 to 3,400, thus reaching engine builders' elusive goal of gaining one horsepower per cubic inch.

The Connie L-1049, built for the Navy in 1956, has four of these turbocompound Wrights and was called "super Connie." This was the model standing before me on this day out at the airport.

It was mostly navy, dark blue-grey, sort of mottled where a new panel had been added here or there. There was a fine smokey coloration to the nacelles just aft of the cowlings, like somebody had been barbecueing in there for a long time. There was not a living soul in sight, but a spindly little ladder of metal rungs was lowered from a rear cargo door on the other side. I walked under the airplane, almost feeling the mighty being above me raise the hair on my head. Still nobody. I carefully ascended the little boarding ladder to poke my head inside. "Hallow the Connie," I called.

"Hallow yerself," came the reply. It was Chuck Wright, the flight engineer. Forty years a flight engineer, mostly in Connies. They say Chuck only tolerates pilots. He said the crew of this one was coming back from town and gave me permission to come aboard.

Gazing from the aft door, down the long narrow tunnel of the Connie, I saw the interior had been stripped. A cargo plane. Here along the Gulf Coast, so near to Mexico, one never asks what big old cargo planes are doing. Bad manners.

I slowly walked forward, admiring the flush fittings built into the deck's end-to-end runners. Graceful tapered, the Connie's fuselage narrowed down at the windshield, leaving room for only the two pilots seated shoulder-to-shoulder. A third station, the flight engineer's, was sidesaddle facing the right just behind the pilot seats.

I just stood there in silent awe, admiring it all. The flight deck was like standing in the workshop of some craftsman who loves his tools and uses them a lot. Wherever hands grip, it was all worn down to bare, bright metal. And 40 years or so sure made a difference in what flight instruments looked like.

Chuck Wright silently came up and stood behind me in the narrow doorway that opens back into the cabin. There was no doubt from his every manner that I was a guest in his home. We were both comfortable with that.

Chuck began to tell me about the Connie. Said he liked the engines, understood them, that they had been used in the DC-7 too, and that engine fires on takeoff were no longer a problem. Then we heard voices and felt movement; it was J.K. West and party coming aboard.

They all gathered up at the narrow end. J.K. West introduced me to Earl Atkins and Atkins' pretty wife, Patty. Earl owns the airplane and flies it. J.K. flies co-pilot with him. West squeezed by and got into his right seat so there would be more room.

"You want to tell me about all this?" I asked Mr. Atkins.

Deceptively mild and soft-spoken, Earl Atkins said he had bought the Connie at an auction of military surplus. "There was nobody else there to bid for it," he added simply. He said he got the million-dollar airplane real cheap but there was nothing cheap about keeping it in the air or finding a place to keep it and somebody willing to insure it. All present nodded in solemn agreement with this.

We got into a detailed discussion as to exactly which model of Connie this was. Lockheed gave a different letter designation for each change they made and a different one for airplanes that went to the Navy. This was a Navy airplane. In time, the Navy would buy a slightly larger, heavier Connie, some with big radar domes stuck all over their outsides. These were used as hurricane hunters; some were to become AWAC loiterers. Some had tip tanks added for range, but none had bigger engines than the one we were standing in. "We sort of have the hot rod of the Connies," grinned J.K.

Takeoff weight used to be 142,000 pounds, but that was with the correct 145-octane avgas which is now all gone. With the available 100-octane fuel, the airplane is restricted to 120,000 pounds. "But we've got range," said J.K. "At low power settings we can stay in the air for a straight 18 hours. That's only 55-percent power, but we still flight plan for 230 knots TAS. And even at that, we are burning off a hundred gallons an hour."

"What did the Navy use this one for?" I asked, while dying to ask what *they* used it for. It had been in med-evac service. All those floor fittings were for the tiered litters, nurses stations, and galleys. The airplane had temperature and light controls all through the cabin. "We fly freight in it," added West.

"What kind of freight?" I asked after a brief pause. That seemed like a friendly enough question, and this seemed like the moment to ask it. They all laughed, knowing my mind. "We fly auto body parts from makers in Mexico to Detroit. Freight forwarders call us. We go out and load up at the border towns of Laredo, Brownsville, or McAllen, which is where Atkins lives and bases the Connie."

"How do you keep separated from the smugglers?" I could ask that now.

Earl Atkins answered, "The U.S. Customs Service. Even though I live there and they know me and know this airplane, you wouldn't believe the ramp checks we get. They look at every part of the airplane; they look at every piece of paper in it. They search us every trip like it was the first time."

Atkins operates more than one air freighter out of McAllen. He and J.K. were in Beaumont now to pick up a Navy R4D, which was the last model of the DC-3 but wasn't flyable yet. They were obviously very proud of the Connie and enjoying her a lot. I asked what the old speedster was like to fly.

"Best airplane I ever flew," said Atkins.

"A dream to fly, does exactly what it is supposed to do," added West.

But Connies in flying condition are getting scarce. They told me of airline crews that come down the ramp to visit and to look at almost every stop. It's been a long time since anybody has seen one.

Atkins said that young-voiced controllers "ask us what it is." And in the quiet traffic at 3:00 A.M. over Little Rock, we had a youthful controller ask us to describe it to him. We did.

And they told of once being on the same ramp with Air Force Two. Of starting out on foot to go over and look at the Vice President's airplane and meeting the crew of Air Force Two halfway. "Can we come look at your airplane?" both crews asked, then started laughing.

Now it was time for me to go. I told them I was going to stand outside and in front so I could hear the music. J.K had his right side window open and was looking back towards the No. 3 engine, whose prop was grudgingly starting to turn in the bitter cold wind. The little APU that was furnishing power to stir that engine in cold oil must have had a heart made of stone. I counted seven blades, then "jerk," "kick," "belch," another "belch," then the blades began to spin, and the sweet music of that giant 18-cylinder Wright flooded the ramp. I started jumping up and down, both from excitement and from being in the cold wind. J.K. looked forward at me; he was laughing, the wind blowing his hair. He had No. 4 slowly turning now. That one began with curls of smoke clouding the lower cowling, then a bright flame, licking. J.K. advanced the throttle, the music rose up into a harmony with the other engine, and the flames blew out.

Now for the two engines waiting on the other side. You could have roasted wienies on the bright flames from No. 2's exhaust. I was still jumping up and down like a kid getting too much excitement. No. 1 began with the majestic turning of cold blades, then started. For a moment the Connie just sat there, all of them turning, the echo of her sweet harmonic bellowing coming back off distant hangar walls.

Then, slowly, the high-legged, droop-nosed Connie began to turn away. Now I know where I had seen that fuselage line before . . . on the Concorde!

From the far end of the airport they lifted off, before the sound of applied takeoff power could reach us. Going away toward a low overcast, a bright flame flickered steadily under No. 3. "Chuck will lean that out," I thought. The fire snuffed out, and the big slender grey bird entered into the clouds.

I phoned J.K. over at Angleton that night about 7:30 to ask how long the takeoff run had been. His wife Shirley said J.K. was already asleep. Why not? What else could a man do with one day? But J.K. called me back early the next morning. Said they were light on fuel, got off in about a thousand feet, and over to Angleton in just a few minutes. Super Connie!

35

Like Torpedoes Being Loaded

IT HAD BEEN A LONG TIME since I had flown my old friend, a Cessna 172. I see lots of them, they are everywhere, sprouting like dandelions from the airport clover. If my pathway along the flight line takes me by one, I always stroke it with a loving hand, saying to myself, "Hello, you fine airplane."

The last one I had flown, over ten years ago, belonged to Bob Marsh over at Houston Hobby. He used it for instrument instruction. It used to sit on his ramp nose-low, like a hog sniffing for acorns. He had every state-of-the art navcom instrument in his panel. Two of each. I used to kid Bob about how long it would be before he would have to install a dual-wheel nose gear to hold up all that weight. But his 172 did what it was supposed to do, and did it real good. It held steady while we students twittered around under the hood, bathing the controls with sweaty hands, trying to keep the needles crossed and centered while learning to fly IFR. On a 172 it was blessedly easy to keep the shiny side up.

Now, on this day I was about to make another trip flying a 172. I had gone into Bob Walker's office at Professional Aviation out at what used to be the Grass Airport before they paved and lighted it and called it, properly, Beaumont Municipal. I explained to Bob that my old Mooney was gone this-a-way and I needed to make a trip that-a-way. Did he have an airplane I could fly?

"Bax, I just don't have anything in right now." Bob was not selling any new airplanes, but the public was sure renting and flying the pants off the ones he had.

I began to finagle. Finding an airplane where there are no airplanes is like finding a hotel room when the desk clerk tells you he has no rooms. Don't ever believe that. A desk clerk always has a room, somewhere. Start off by offering to sleep in the broom closet, or the bridal suite. Make no signs of leaving.

Soon Walker admitted, "Well, of course I have the 172, but you wouldn't want to fly *that*."

And I said, "Bob, compared to making this trip all day at 55 miles per hour in my car, I would dance at your wedding for a chance to fly a 152, much less a 172."

Bob sort of halfway grinned, said thanks but he was already married (as I very well knew), and then turned his gaze toward his instructor/pilot James Joynt. James sort of sighed and nodded. He was trapped in the office again when a lousy flying deal came up. One of the most important things for the company pilot to remember is not to get into the back of the office—where you can't sneak out—when you hear the conversation working its way around to some kind of a sorry flying assignment. Hang loose near the door so you can vanish before they think of you.

And so it came to pass that I was back at the handles of a Cessna 172 again, after all those years, and James Joynt was stuck with making the trip with me and ferrying the airplane back.

In the roomy cabin of the pretty white bird with the burgundy-and-gold stripes, all the old memories began to come back to me. I still knew where everything was on a 172, partly because Cessna had been kind enough to not switch stuff around very much in the past 10 years.

Or very much in the past 30 years either. The sweet memories of my first 172 flight came flooding back from 30 years ago, when the 172 was a new and strange airplane at small-town airports. I got to fly the 172 by default. All the two-seater taildraggers I had been flying around in were used up or not available, so Van Vannerman and I had marched out to his brand-new 172. He would rather risk me flying his new plane than lose the fee of an hour's worth of dual time.

I flew it briefly, then flew my private pilot check ride in it, too.

Compared to the Aeroncas, Luscombes, and Cubs I was used to, that 172 looked as big as a DC-3. But Van calmed me with words I have heard many times since. "It's just another airplane."

I had never flown anything with a nosewheel, flaps, or a trim wheel. A four-seater. I was just sure it was going to eat me up. Those first 172s were mostly unpainted—bare gleaming aluminum, square-backed, squared-off rudder. In the pilot's seat I seemed to be way up high, but I quickly learned that the 172 was easier to fly than those nervous little taildraggers. The only thing I had trouble with was its blinding speed. Everything kept coming up about twice as fast as it used to. For a while the 172 stayed ahead of me, but I passed the private pilot exam. And in all the good years to come, I flew the 172 whenever I had the choice. But when the time came to buy an airplane, I opted for more speed. Couldn't afford a 182, so I bought a Mooney.

Now, after all those years in the little round cabin, I was strapping on another Cessna, sitting up in the big square cabin, enjoying the rush of good memories. James Joynt and I planned the trip. I told him my usual route to Kerrville was north of Houston and south of Austin. In pretty weather Joynt saved a little time and money by cutting across Houston Intercontinental. I didn't know you could do that. But that's what he requested, and that's what ATC gave him: direct to the Humble VOR, which sits right on the airport, maintain eight thousand five hundred, and they gave us a squawk code. You couldn't ask for anything nicer than that. I suppose a medium altitude over a big jetport is the most out-of-the-way place a lightplane can be.

I looked forward to the eager way a 172 develops lift, gets off, and is gone. After settling down to cruise, I realized I had almost forgotten the sounds of a 172—the constant whistling moan of the struts and the way the prop will cavitate and snore in rough air.

Trimmed and level at eight and a half, we were nearing Intercontinental. Joynt was working the radio; I was just having fun. He leaned over toward me and said, "Look out the side." Right outside my window, close enough to read all the lettering on the fuselage, was a big, fat, shiny 727, ghosting past us in a slow descent. Beautiful!

As we cruised level inside Houston's transition and descent area, ATC was passing 727s around us. They started passing us at about two-minute intervals, mostly Boeing 727s, a few Douglas DC-9s. Man, talk about holding altitude and VOR heading like it was some kind of a religion! One of them seemed to level-off momentarily when it got right beside us, as if to give all the passengers a good look. We looked back.

I could just imagine that co-pilot on his cabin speaker, "See that little airplane just off our right wing? That's how it is supposed to be done. Approach Control told us where to expect him, and he is right where he is supposed to be. There is no danger involved, not even as much as passing a slower car on the highway . . . pretty isn't he? That is a Cessna 172. Both the Captain and myself learned to fly in an airplane exactly like that. The Cessna 172 is to general aviation what this Boeing 727 is to the airlines. The bread-and-butter airplane."

At least that is what I was hoping he said.

And still they came, a steady parade of colors. Watching each one go by made the bottoms of my feet tingle. They slid by like torpedoes being loaded soundlessly.

Then the situation down on the airport changed. A just-landed airliner lost hydraulics on the active runway, and Houston smoothly rearranged their landing pattern. The last one was even closer than the others; he was crossing over the top of us.

"He see us?" I asked Joynt.

"Yeah, I heard him call us in. I think he's doing the crossover just to get lined up on their new landing pattern."

And sure enough, he reappeared in the top corner of the windshield, going away. Nothing but those three black holes. Sure looked business-like.

He passed high enough to avoid rocking our boat, then my VOR needle went into its zone of confusion. We were passing the VOR. On the other side, outbound, there was no traffic.

I took a deep breath, realizing I had been gripping the yoke. "Wasn't all that a pretty sight?" I asked Joynt.

"Best way to see one, air-to-air," he replied.

An easy silence settled over James and me, each of us with visual memories of those airliners in flight. We flew on awhile in silence, then I said, "With any luck at all the next town up there will be San Marcus." It was. Pretty place. Kerrville and the hill country would be next. The land rose up beneath us; the clouds began to build up overhead. We found a long open slot and went below the cloud base, back down into the heat and bumps again, expecting Kerrville just over the next ridge. You ever notice how the last twenty minutes of a trip can seem to last forever if you are too anxious to get there?

Schriener Field of Kerrville lies folded into a valley of low, limestone, mesquite-covered ridges called the Texas hill country. Sometimes you can see deer, frozen at the sound or shadow of your passing wings. To many this hill country is the most desirable part of Texas to live in. Real estate prices have gone steadily upwards as more winding trails snake up the hillsides to little villas on top. The air is mostly cool and dry. You could shake out a dozen retired Army colonels per mile down there. And beyond each ridge I thought I saw the clean, long, low, sheet-metal roofs of the Mooney factory. Mirages to eager eyes.

One more ridge, and there it was for real. I entered downwind. Joynt softly said, "Too high." It looked okay to me, but I came back on the power and lowered the nose. "Too fast," said Joynt.

What did he want, a hover? The approach still looked okay to my Mooney eyes.

"You won't make it," Joynt said, almost to himself. And suddenly there was the threshold, and the 172 seemed to be three miles high and floating like a leaf. I decided I was too rusty to try any of the good stuff like a steep slip or straight-down shuddering and shaking under full flaps. Instead, I just let it come on in.

We never got near the runway. Joynt picked up the mike, said to the temporary tower, "Missed approach."

Tower, having a little fun at our expense, replied, "So we see."

I applied power, got out of my Mooney landing suit, and tried to recall the gentle approach and landing speeds of a 172. The landing was, as they say, uneventful.

I had hoped nobody had seen us, but there they were, lined up like cawing railbirds at Dugosh's gas pump. "Hiya, Bax. Where'd you get a Cessna?" "Was that you trying to land out there?" "Bax, were you the instructor or the student?" "Haw, haw, Hawww."

It was bad enough to arrive at the annual Mooney Pilots Homecoming in a Cessna, but to miss the whole damn airport upon arrival can last a man for days.

36

Testing the Tires

THE TROUBLE WITH OWNING ONE AIRPLANE for a long time is that the airplane becomes a part of the family, and you tend to use it only when you need it. That is when you find out that, while you have not forgotten how to fly, there are lots of things about good flying you have forgotten.

I just came in from a session of touch-and-go landings, and my sweat is still drying. I'll bet I made more landings and takeoffs this afternoon than I did in all of last year. And that is what I needed.

I went out to Beaumont Municipal with the intent only on finding somebody to fly the pattern with me. To do what the Brits beautifully call "bumps and circuits." I got a big howdy from Bob Walker, FBO, the keeper of our Mooney, and told him of my need for a pilot instructor. His gaze once again fell upon the luckless James Joynt. James was trapped in the back of the office again, in the big easy chair, soaking up the air conditioning after a hot day of dual. Joynt let out a little moan, nodded his head, and heaved himself to his feet.

Once again, if Joynt had known to not hang around the front office he'd be on his way, home free, instead of about to share another hour or so of low-level, hot, sweaty work with me.

I have gone through a funny transition during my many years as a pilot. When I began, thirty years ago, I was the kid, and all the instructors were old men of thirty or so. Then, suddenly, it seems I am the old man and the CFIs are all kids. I don't remember if there was a time when we were all about the same age. Life will do that to *you* too, if it hasn't already.

James Joynt is one of the new generation of aviators who got licensed knowing more than we did after years of flying. He is fair of hair, blue of eye, smooth of brow, a darned-good CFI, and as I had previously learned, not one to chatter. But his few words should be heeded. As they say in my native deep east Texas, "If he says a hen dips snuff, you can look under her wing for the box."

On our walk out to the T-hangar, way down at the far end where faithful old 1968 Mooney Ranger N6727N has lived since she was new, I explained to Joynt what I intended to do aloft. When I'm about to embark on a flight with somebody else I always appreciate it very much if they outline the intended flight syllabus. I don't like surprises. Especially in airplanes. And one of the hazards of my trade, that is, flying and writing about it, is being strapped in with some gung-ho company demo pilot who wants to take me up in this pre-production prototype and show me all the hair-raising things the ship will do. At times this leads to some awkward moments. When a new aircraft starts doing things that surprise both of us, I just feel good to be able to get out of the thing on the ground and walk away from it.

I always tell them, "No funny stuff. Just take her straight up and straight down."

On this trip I explained to Joynt that all I intended to do was stay in the pattern and learn to land my own airplane all over again. My work, while keeping me airborne hither and yon, is most sadly lacking in the basic practice of landing techniques. And this was what I wanted to work on.

I told him I would land in all flap configurations and do some short field approaches, too. Joynt solemnly agreed to the wisdom of this, revealing that it would be his second right-seat workout in ol' 27 November this day.

It seems that my partner, Elmer Lee Ashcraft, a CFII, had been given formal written notice by our insurance company that, if he did not complete a check ride by a certain date, our insurance would be null, and as they say, void.

The idiotic part of all this was that my partner did not get the threatening letter until past the date he could do anything about it.

One of these days, soon, we ought to all sit down and have a nice little fatherly talk with all these aircraft insurance writers.

Anyway, our airplane was still warm, and Joynt knew his way around the cockpit very well. For that I am still thankful.

We got the airplane before the fuel had been topped off from Elmer's check ride, and one of the linemen advised me to expect to find some fuel usage when we looked into the tanks. Now that's what I call professional. They also know we never let the sun go down on our Mooney without both tanks being full. We think this may be the reason for the long, trouble-free life of the sealant in our tanks and the small amount of water we find in preflight drainage. This could be just an old Mooney's tale, but it has worked well for us.

The line crew also drew the airplane from the hangar by tow bar, for which I thanked them, although I was already in the cockpit and had been planning to blast it out from the hangar under its own power. This is my own private drama, the airplane coming awake inside the hangar and emerging with its prop gleaming. This always seems sort of grand to me.

The homeplate crew drew the airplane forth by hand, knowing well that we do not permit the use of a tractor on the other end of the towbar—not there nor anywhere else. This is another of those little Mooney idiosyncrasies we learned of the hard way long ago. The nosewheel of the Mooney has a very restricted turning arc. A big strong lad riding a tractor is capable of trying to turn the airplane too sharply; the towbar then dents the nosewheel truss. The cure is replacement, and the price is more than you would imagine. So Mooney owners tend to hang around on the ramp and guard their precious craft against anything but hand towing.

Now, with everyone out of the way, I shouted "Clear!" and brought that faithful 180-hp Lycoming to life. Hearing that old love song again, I felt oddly content and at home. But I cautioned myself against too much daydreaming. Letting down your guard leads to mistakes. So I went down the checklist as if I had never before seen our old airplane.

My intentions were to show this fine lad beside me a flawless flight. Ha! There is nothing like an airplane to humble a man.

When I slewed the tail side-to-side on taxi-out, I explained to him that I was checking the response of the needle and ball, which was true. And after the runup I asked him if I had forgotten anything.

No? Then let's go.

We have earmuff-style headsets to protect our hearing from the Mooney's song, but in dual work I ride with my right ear uncovered, both to hear Joynt and to hear the natural sounds of the airplane. After all these years of flying I still fly somewhat by the sound of the wind in the rigging, although it's been a long time since I had an airplane with any outside rigging.

I go to full takeoff power in slightly delayed stages. No need to cram it to the engine on first takeoff when we have all that room ahead. Oh Bax, you old fox.

The first approach and landing was what I do mostly at big airports where they often ask me to keep my speed up because of fast following traffic. Again I didn't look at the airspeed indicator but flew by sound and feel, as I always have in such landings. No flaps, no flare, just holding her off the runway, constantly nibbling at the controls until the main gear squeaked. Not a bad landing. A high-traffic, long-runway, metro-airport arrival—although the nosewheel wanted to veer around after its high-speed touchdown.

The next landing was announced as half flaps. There are those who believe that half flaps, or any amount of flaps, on a Mooney is only academic. That the slick little speedster is going to land the same no matter what you do with the flaps. Not true. The delusion is caused in these older models by the lack of pitch change at any gear or flap position. The hidden genius of Al Mooney peeks out at you. But all of that ended with the six-cylinder engine on later Mooneys, which added 10 inches to the nose.

And so I was coming in with half flaps and at a still-too-fast 90 IAS. The Mooney got close to the runway, but that's all. It went into ground effect and began to demonstrate its famous total float, which it will do, just off the runway, until it gets down to its normal touchdown speed. Way down the runway.

I reached over and sucked up the flaps. Not a good practice for learners, but the retraction of flaps matches the sink rate, and the result is a much shortened gentle touchdown. On a newer model with power landing gear, you could just as easily touch the wrong switch and suck up the wheels. This results in an even shorter —and noisier— landing.

"Full flaps next," I announced as we went round the pattern, still too fast and too close-in. Joynt nodded, then reached over and pointed at the red "unlocked" warning light on the landing gear indicator. On these old manual-gear airplanes it is possible to have the gear up and seemingly latched, but if it is not latched smartly enough, it will not close the microswitch inside the latch, and you won't get a green light. On a cross-country, with the gear handle latched safely to the floor, you can stick a finger down inside the gear-down latch (on the bottom edge of the instrument panel) and feel that little microswitch with your finger. It won't bite.

I thanked Joynt, unlatched the gear handle, moved it very slightly back, slammed it smartly into place, and got the green eye.

I am thankful none of you saw this hot-day, too-much-speed, full-flaps attempt at landing. I was, however, able to demonstrate that the prompt and judicious application of power will bring a Mooney back down from high funny places. Joynt's hands fluttered in his lap, but he was kind enough, or brave enough, to let me complete the struggle alone. Fine young man.

"What did I do?" I asked on the go-round.

"Too fast," he replied with heroic restraint.

I decided to cool off some by just making a high-speed low pass down the runway just for the fun of it. Low as I could go and not do more than roll the wheels.

"You gonna do this with the gear up?" asked Joynt as we came sizzling in on final.

"Oops." I lowered the wheels and made a nice screaming pass, which is legal by the way.

This time I doubled the size of the pattern I had been flying and gave up those carrier deck approaches. Voices of long-gone instructors came back to me.

"Bax, you cannot expect to make a good landing from a bad approach."

"Bax, you have excellent recovery from odd attitudes at low altitudes . . . now let's just fly some good old standard stuff like it says in the book."

"Bax, I'm not really teaching you anything until I get your shirt stuck to your back."

I settled down to watching the instruments some, instead of trying to fly a Mooney like a Jenny. With over-the-fence speed reduced to 70 from 90, things settled down a lot.

"What airspeed did my partner cross the fence at?" I asked.

"Same as you are doing this time."

At 70 the airplane had more flare control, touched down slower, and was not so squirrelly on the roll out. And my shirt was plenty stuck to my back.

We did some more of these tame ones, got some chirpers, and when I said, "Let's make the next one a full stop—if I still know how to do it—and head for the barn," Joynt did not plead for more. We taxied in with the door held open, cooling the cabin. "Pretty bad, eh?" I asked.

"Pretty rusty at first, but you were settling down." This man could work as a diplomat.

We taxied to the ramp, me feeling on my high keys as always after any flight, good or bad. In the terminal building we met an old friend in the corridor. "Where you been, Bax?" he asked cordially.

I told him we had been testing the tires. Joynt was kind enough to turn his head away to hide his silent laughter.

37

A Day in the Life of a Girl Watcher

As a conscientious and practicing girl watcher and biplane ogler, I always enjoy airshows. They provide a plentitude of both biplanes and pretty girls. Each species has come to see and be seen, and are looking their very best. There is seldom any lack of subjects. If there is no pretty girl passing by, then there is a hoary old biplane to see, standing tall in its classic splendor. The hours spent by me at airshows glide by swiftly, as time does when one is happy.

But last summer (1985) I encountered a situation of memorable split attention; there before me was one of the great biplanes of aviation history with an unbelievably beautiful girl in it.

One of the natural hazards of pretty girl watching is to be regarded by the subject as an Old Fool. Fair enough. This pretty lass seemed to be on duty with the biplane; she was engaging the passing public in bright and knowledgeable conversation about the old airplane. Well, that happens sometimes. The company that owns the airplane will hire models, give them a skimpy background on the airplane exhibit, and have them stand there and pose, be friendly, and hand out company brochures. No harm in that. Good crowd attraction, too.

I had already recognized the biplane. It seemed to be a flawless replica of the rare and famed Travel Air D4D, a few of which were built in 1929 at Wichita by the Travel Air Airplane Company which at that time was bursting with so much talent that the major partners of the company were about to split up with independent designs and different companies of their own. Walter Beech would remain, changing the name of the company to Beechcraft. Clyde Cessna would begin his own company based on a clean monoplane design that he believed in. Lloyd Stearman would continue to build outstanding open-cockpit biplanes, but under his own name. And another tall young man, not in the company, although invited, would continue his search for a small high-performance monoplane with retractable gear. His name was Al Mooney. What a time in history to have been in Wichita.

The Travel Air D4D was the high water mark in open-cockpit biplane design. It featured a then-radical airfoil: a very thin wing, built for speed. The rest of the airplane was clean too; it had streamlined wheel pants and one of the first fully cowled engines. This airplane was designed as a racer and won its share of them too with "ol' steely eyed Walter at the controls," as Al Mooney described it. This big-engined Travel Air beat out one of Al Mooney's advanced Eagle Rock biplanes and laid Al's biplane designs to rest forever.

I knew the Pepsi Cola Company had bought one of the limited-production D4Ds, painted it up in gaudy Pepsi colors, and it flew it long and hard and with great success into the nation's awareness as the famed Pepsi skywriter. Whoever had done the replica that I was now admiring had done an uncommonly accurate job of it, exact in every detail.

I struck up a conversation with the pretty girl, of course. She looked like Shirley Temple should have turned out to look. She could have been a movie star.

She said her name was Suzanne. That's nice. I think she gave me her last name too, but we were at the EAA airshow at Oshkosh and a squadron of AT-6s were passing over in formation just then, so I was just watching her face, not being able to hear a word she was saying. But enjoying her animated manner of speech. There also seemed to be a friendly brown beagle as a part of all this. The dog was sitting atop the fuselage as if she owned it. The dog's name was Charlie Brown. Leave it to the Pepsi people to think of everything at an airshow display. A great old airplane, a pretty girl, even a friendly mutt by the name of Charlie Brown.

I was trying to hear Suzanne and at the same time get out to the lower wingtip of the airplane so I could sight down it and see if this great replica had the same thin wing-rib section as the original had.

It did. I wondered how they got the pattern for it.

My attention back to Suzanne standing by the cockpit, I noticed we were both being calmly watched by a fine broth of a lad, all six-foot-two of him. "Husband." My husband alarm is set very fine; it always goes off on time. I watched the young man very carefully. He was a pilot. No doubt about that. From his air of easy familiarity I guessed he was the pilot-mechanic of this show plane. That was the only guess I got right all day long.

I grinned, nodded, and stepped back a pace to include the young man in our conversation. Always grin, nod, and step back a pace in the event of sudden husband.

I decided to just worship Suzanne from afar and include the dog in our conversation as a safety backup. The nice old man likes dogs, see?

Charlie Brown turned out to be the mascot of this flying team. "We take her with us. Nine years old and has logged over 800 hours of flying time in the plane. Loves it. We even have a beagle helmet and goggles for her."

"Do you fly the plane?" I asked Steve Asbury-Oliver as Suzanne introduced us. Steve nodded that he does. Not much of a man to carry on with strangers.

Suzanne, friendly enough for both of us, was filling in the gaps in the conversation and enjoying it. "He's my mechanic and pilot, too. He keeps old Nancy in immaculate condition, ferries her to the airshows. I do the sky writing."

I gulped and swallowed a time or two. I was still trying to get all this to compute. "Nancy was the name of the original Pepsi sky-writer back in the 1930s," I offered.

"We know," said Suzanne, eyes shining brightly. "This *is* Nancy, the original Pepsi plane. They once had six of these Travel Air D4Ds."

"You mean this one is just like Nancy?" I flip-flopped.

"No, Mr. Baxter," said Suzanne Asbury-Oliver as though explaining something to a dense child, "This *is* Nancy. One of the originals. This is the airplane that first wrote 'Drink Pepsi Cola' in the sky over New York City on May 1st, 1938."

"She's the pilot," offered her husband. "She's an ATP, started with gliders back home in Forest Grove, Oregon. That's near Portland. She soloed at 15, got her commercial ticket at 18, and became a CFI.

Taught lots of old men to fly, including her own Dad," said Steve, looking me straight in the eye.

He continued about the airplane I had mistaken for a replica. "The D4D was the fastest non-military biplane of its time, had a 350-hp Wright Whirlwind engine, service ceiling of 20,000 feet. The recent restoration of this original airplane cost $100,000. Took 8,000 manhours of work."

"How did you happen to, ah, meet Suzanne?" I asked.

"When Pepsi decided to go back into skywriting she was one of over 3,000 applicants for the job. Pepsi was doing this to celebrate its own 75th anniversary. Smilin' Jack Strayer, a veteran of 40 years in aviation, took the plane on tour for Pepsi and took Suzanne as co-pilot and trainee. This was in 1980. Strayer retired the following year, and Suzanne took over the flying. I met her in 1981 at the Kentucky Derby. She was skywriting for Pepsi; I was towing banners in a Stearman. She just swept me off my feet," said Steve with a grin.

"I taught him skywriting, too," said Suzanne. "He soloed on Valentine's Day. Six days before we were married he went up and wrote three perfect Pepsis, then painted a great big heart in the sky for me."

Suzanne then went on to tell me that Steve, originally from Rutledge, Missouri, was also an ATP, had gotten *his* commercial and CFI at the age of 18, and is also a certified controller and radar operator. As man and wife and beagle they work the skies over about 160 cities a year.

"Tell me, how do you do it?" I asked.

"I work at about 14,000 feet," said Suzanne. "The letters I smoke into the sky are about a mile long. A good sign is visible for about 1600 square miles. The writing itself is like flying a dance routine. Timing is essential. So is concentration. I write in mirror image because my readers are on the ground looking up."

"Well, you sure can't come back and erase any of it," I offered. "Have you ever goofed?"

Suzanne laughed and was beautiful. "Yes. Once over Chicago, Center advised me that a 747 was coming my way at my altitude. Watching for the airliner, attention split, I didn't realize until I had gotten back onto the ground and looked up that I had just invited all of Chicago to drink 'P-P-E-P-S-I.'" Now Suzanne was laughing at the memory.

Standing there, gazing upon man, dog, wife, and their historic old Travel Air, I told them I thought they were W-O-N-D-E-R-F-U-L!

And Steve? He said, "The first time I got into that old Pepsi biplane I had the strangest of feelings, knowing this was the very same airplane I had seen in the sky as a boy over 30 years ago. To me it's like a dream come true; flying the plane I love with the woman I love."

After this story was printed in *FLYING*, Suzanne wrote me a special letter. It said simply "T-T-h-a-n-k-s!" Such fine people. May their shadow never decrease.

38

Grounded

FLYING MAGAZINE IS WRITTEN BY PILOTS, a round table of specialists having in common our love of the sky and our joy of writing about it. There is the close knit little group of full-timers at the New York office, but almost as many of us, the "contributors," live and write from far afield all across this country.

We are our worst critics. Before a story is published it goes "around the chairs." No writer should ever see what marginal comments his prize story might gather, but this is also where we search out historical or technical flaws in each other's work. There are close friendships, mutual affections that flourish under the editorship of Dick Collins, who came up through the ranks. One of those writers magazines was once asked how you get into FLYING. They replied, "You have to be born there."

As is common in the world of freelance writers, most of us "contributors" do something else for a living. I have been a Texas radio and tv broadcaster for over 40 years, but the real joy and glory in my work has been writing for *FLYING* since 1970. To go out and fly airplanes and then be paid to write about it still sounds like something I might have dreamed up.

It was never what you could call work. I would go out and fly and come home with the story full born in my soul. I just sat down and let it pour out, never needed a re-write, and the magazine seldom changed a word of it. Then, 10 years ago, I lost my medical. Lost it for good. I have an undiagnosed condition of small blackouts. Like somebody was playing with my master switch. I've had all those expensive tests of waves and pictures of my brain, and one good ol' doc even offered to open my skull and take a peek inside. I told him I'd play the hand I got.

"We are the generation who play the hand we were dealt," said old friend Gerald Fay, retired DC-10 skipper and now a published writer. Gerald is one of those still-young men who took his mandatory retirement at 60 and walked down off the flight deck with tears in his eyes. Can you name any other form of human endeavor where a person can put in a lifetime working at it and then retire in tears?

I wept when I lost that medical, but didn't make a case with the feds about it. The FAA is real picky about wanting the pilot to be fully awake and in possession of all his marbles at all times. I just let my medical quietly expire. If the conditions ever change, and they won't, I could go back and sit for a new medical with only a long lapse of time to show for it. I don't want the feds to have any record of mine to chew over.

I wrote about it when I lost the medical. I never said why, and got a flood of letters from you, most thinking I had a heart attack at 54, right on time. In truth I always exercised, am a lean, wiry 64 now, and the doc says "You've got a cardiovascular system like an 18 year old." But I saved all those letters. Beautiful.

And I began to try to find some rat hole to keep on flying. I had just bought our Mooney a year or so before, and for the first time was free to get out and do a story, get back again, and not miss my daily morning radio shows. Can't miss that. Radio feeds the writer. I simply could not accept the fact that I was grounded. Flying had meant too much to me all my life; the getting there and all those ratings were too much for me to be able to turn a page and say it was over. I simply refused to recognize the truth that I was grounded.

Lots of my old hangar buddies said, "Bootleg it. Who ever asked to see your medical?" Collins said he'd fire me for that, and my wife Diane said she'd quit me. For a brief time I had a secret affair with my airplane in stolen solos. I sold partnerships in the Mooney so I could keep it, and my hidden set of keys.

Then one clear cool night I was IFR solo, out of Houston to homeplate Beaumont. I believe in safety and always filed IFR, even on a clear night like this one. It was one of those velvet nights. I had the old Mooney trimmed to a gnat's eyebrow and in easy cruise. I looked down at what I thought were the lights of Baytown and wondered why Departure had not handed me off to Center yet as they always do about there. I gave them a call.

Departure came down on me like a falling brick wall. "Two seven November, where you've been? We've been calling you for 20 minutes!" and on and on.

I looked down again. That was not Baytown; it was Winnie, a long ways past Baytown. Realizing I had been out of it, I made a fast scan of the panel. Everything was just like I saw it last—heading, altitude, all of it. Thank you, Mr. Mooney.

I told Departure I'd had radio trouble.

When I reported Beaumont Municipal in sight to Beaumont Approach, I suddenly realized an awful uncertainty about landing on this unlighted home field. I'd done it for years, but getting all my wits back was a slow process. I began to talk myself down like I was a slow-witted student. I could do what I told me to do. Shaken, and alone at the airport that night, I pushed the Mooney back into our hangar and never flew solo again.

But I still could not, would not, admit to me that I was grounded. I found another rat hole. I would go out to the airport when I had a trip to make and ask, "Who wants to log some pilot-in-command time in a Mooney for free?" That was a bargain to whomever was there. Not the best of flying, not like being alone with the sky and one's innermost thoughts, but it was still flying. (If you still fly solo, cherish every minute of it.) I carried my license in my pocket. In my own mind I was still a pilot. The writing was good as ever.

Then this year our insurance company ruled that all pilots in that Mooney partnership had to be named pilots on the written policy. My partners didn't have the heart to tell me; I blundered into it myself. And that was my autumn of grief. My last rat hole closed. I put my partnership in that good and faithful old Mooney of so many sunny miles and a few wild storms up for sale. I never went back to look at her. I also believe in closed coffins at funerals. I didn't want to remember her cold.

It kept coming over me in waves, "Not a pilot any more. You never were. You've been a living fool for the last 10 years." I wept bitter tears.

Diane put her arms around me, "I wish there was something I could do."

"There was. You could have gone on and got your license and we could still be making trips together."

Cruel. I felt cruel. And for the first time Diane leveled with me.

"I'm scared." she said.

A lie. She has 8,000 hours of crew time in 10 years with Braniff and has seen every cabin emergency. You can't scare Diane around airplanes. But she is also the late-life mother of our only most beloved only-child, Jenny. Diane changed from a sky goddess to a she-bear. I can't blame her.

Not a pilot anymore. We've never done a story about that. Think of all the others it must happen to; grounded is a part of flying. I had been hiding, covering it up. I never told Collins or anyone at the magazine. I didn't need to. My writing changed. For the first time in my life I got three stories in a row rejected. They were real polite, like being around an old pitcher who has lost his arm. Nobody asked anything.

I tried writing out of the caves of my mind. Thirty years a pilot is a lot of memories. Only trouble was I had already written all the good stories when they happened. I'm in the strange predicament of Baxter trying to write around Baxter, and he's a tough ol' boy to beat.

Then yesterday Collins phoned from New York and suggested that last story I wrote for him could maybe stand a little re-write. Collins is not a raise-hell editor, but the more polite he is, the more trouble you are in. And suddenly I was sick of faking it with Collins, and with you too. I found myself telling Dick all this over the phone, washing clean, feeling better. I told him the only thing worse than being grounded is the thought of not being a writer for *FLYING* anymore.

Collins was calm, even gentle, and made a suggestion I'd never even thought of because it's so obvious. "Why don't you just rent an airplane with a pilot when you need to get some flying done?"

I had even gone back to building models again, holding them up to look at them like when I was a boy. But now you know, and all the doors are not closed. The writing will flow easy again now that I'm not living a lie. And this might be the first good one—about what happens to a pilot when he does not outlive his license. It's never been talked about much. Now you can, with me. That should be of some help, for being grounded is a part of flying as much as death is a part of living.

39

Epilogue

TWO EVENTS IN THIS LONG HAPPY LIFE of flying, writing about it, and being paid for it have stuck in my mind. Both were ideas—not events, not things—but ideas, man's most forceful tools.

The first happened at one of the all-night brainstorming sessions we used to have, with all the writers and all the editors in one room, ideas flashing amongst us like bright swords. This took place about ten years ago, at a time when private aviation and airplane production and new-start students were at an all-time high.

One of our editors stood up and came off the wall with this, "If all of general aviation—pilots, planes, publications, all of us—were suddenly stricken from this earth, the world would little suffer for it nor long remember we were here."

That may not be an exact quote, but it is the exact idea. His words salvoed a storm of replies from all of us, each of us saying in our own view the importance of general aviation. He never bothered to counter us but let his dreadful proposition simply stand as he had said it. The kind of a thought that becomes a burr under the blanket on even the most tranquil nights. What if he was right about that?

An avant-garde thinker, this man. And a close friend. He confided to me he was becoming bored with boring holes in the sky and writing about it. He also admitted to having turned down an offer to be editor of Penthouse Magazine for more pay than he was now earning. "What if I got to feeling about Penthouse the way I now feel from so much aviation?" He really had a delightful, if blasé, outlook on life and its possibilities.

In the last few years, as production of Cessna, Piper, and Beechcraft single-engine piston-powered lightplanes has almost ground to a halt, I have often thought of his "what if" about general aviation. To my mind this would be like having GM, Ford, and Chrysler suddenly quit making cars.

It's really unthinkable. Or is it? The world has kept right on turning without us, as he said it would.

What happened to lightplane sales is still in the jury room. Much sound and fury, but no real verdict yet. Lots of finger-pointing but real villian, or villians, proven guilty.

But I have noticed a strange thing taking place among us. The love of flying, the interest in flying, the attendance at airshows, readership of aviation books and magazines—all of this continues unabated. Growing numbers of Americans are still captivated by the idea of flight, even if all they do is read about it.

How can that be explained? Rutan, Yeager, and the flight of the Voyager most surely had much to do with it. In the true tradition of Lindbergh they conceived of the idea, raised the money, designed and built the aircraft, and flew it themselves at great personal risk. Capture the hearts and minds of all America they most surely did. And as in the post-Lindbergh period, there has been a spate of daring post-Voyager flights.

In the small world of general aviation, with few new airplanes to choose from, the demand has been felt in the used airplane market. And homebuilt projects are no longer sneered at by outsiders. In fairness and fun, I must say that the former editor, who made such a damning observation of us, has long been a very successful freelancer who, in his spare time, is nearing completion of a very sophisticated and popular homebuilt. Beautiful!

As to how and why general aviation continues to hang on, at least in peoples' minds, I fall back to an old and original observation of mine on flying. Mankind has always dreamed of flying. And flying is one of the two human experiences that, once attained, always leaves the desire for more.

I remain convinced that, after being purged through this period of low activity, general aviation—that is to say, the airplanes for the common man's personal use—will emerge in some long-overdue new form and fashion. Here and now, this period of waiting, is not an unexciting time in aviation history.

The second event that touched my mind and shaped my personal thinking about aviation was a book recently written by former *FLYING* editor Dick Collins, titled *Air Crashes: What Went Wrong, Why and What Can Be Done About It*.

In his book Collins opens a new school of thought on air crashes; that they are caused by wrong decisions made somewhere in the chain of events that lead up to the actual crash. Oddly enough, a year earlier the Navy safety publication *APPROACH* had published a feature on this same concept.

To reduce this concept to its most simple and provable terms you only have to look at the DOT statistics on general aviation crashes. For now, these numbers show private flying to be more dangerous than driving your car. However, if all the "wrong-decision wrecks" were removed from the statistical pile (e.g., airplanes that were flown into the ground by VFR pilots in instrument weather; fuel starvation; drunk pilots), then the numbers would be far below those of common everyday car wrecks. Flying *could* be safer than driving, and it *should* be.

Wrong decisions can be made in repair shops, too, but the results are the same. Aviation mechanics are an inseparable part of aviation and want to be thought of as such. But a wrong decision on the workbench can be just as deadly as a wrong decision about scud running. Fatal engine and airframe failures are rare. And if you wanted to carry the "wrong decision" concept to its logical extreme, you could include failed crankshafts, pistons, controls, and riggings from the factories. Even there somebody can make a wrong design or manufacturing decision.

But it's the pilot who is most often and most obviously involved in the decision making in flying. Think back over your most memorable hangar tales, of "there I was . . . ," and find that sneaky always-present culprit . . . your own wrong decision, Ace.

Such thinking is not intended to cast a pall over the pleasure of flying. Airplanes are beautiful, and flying should still have an element of fun in it, but as Collins says, a pilot should never lose his native and healthy suspicion that these things can kill.

Looking for a lighter note to leave you with, there is always that wonderful storehouse of natural aviation humor. Like, what is a "coon-dog" landing? One that goes "Yelp! Yelp! Yelp!" And a bachelor pilot is a guy who comes to the airport from a different direction each morning. Or the Aggie Parachute—it opens on impact.

Or the true story of the commuter line flying Swearingens, which they called "weedeaters." Losing an engine on takeoff, the co-pilot went on the PA system to calm the passengers and said, "Don't worry folks, this engine is perfectly capable of flying on one airplane."

Or the time when my Diane was flying as cabin crew for Braniff. It was still a small family of an airline then, and Di was home sick with what we then called the "Asian flu." The dispatcher called and begged her to take an imminent flight. She assured him she was too sick to work.

"Just suit up and ride the aft jump seat so we'll be legal. All we need is your warm body."

Diane dragged out for the trip, and wouldn't you just know that a Joe in the aft aisle seat wanted her to get up and fix him a drink. Diane tried to explain that the other girls would soon be there and that she was only deadheading as a passenger.

The guy got sarcastic and made the mistake of snapping his fingers under her nose. Diane took his order for a Marguarita, went to the midships galley, mixed his drink, then licked the rim of the glass inside and out before dipping it into the salt.

"Whatever kind of a day he had planned in New York, he didn't make it. He had one of the worst cases of Asian flu ever known."

Don't ever get smart with the stewardi. I, too, learned it the hard way. On a ground delay the stewardess came by, and I said, "Ma'am, I just saw a man go under the wing of this plane with a screwdriver."

"That was no screwdriver," she shot back, smiling, "that was a martini."

And from an old instructor I picked up the habit of always giving each propeller blade, wingtip, and tail a firm, friendly shake. It looked so silly that one time I asked him about this preflight practice and he said, "I've never found a loose one yet, but if there is, I want to be the first to know it."

Then he spent a lot of time fussing with the seat adjustment, ending the ceremony with a firm both-feet shove against the seat travel latch. "First I get me where I'm supposed to be, then I make sure the seat is going to stay there." I adopted that habit of his, too.

So many good airplanes have brought so much pleasure for so many years of my life that I always stop, watch, and listen when one goes over, and try to guess what it is—as I have since childhood.

And I like to look up and watch the changing sky at sunset on a pretty summer day—already dark down here on earth but with rounded summer clouds still up in the sun's light. Taking on the purple of night at their bottoms while their summits glow in the reds and golds of the setting sun. Once in a while I will see a tiny airplane up there, silent, almost motionless, still silver in sunlight; and I think, "We have been there . . . with a little luck will be there again . . . thank ye Lord."

Flying to me, more than anything else, is an experience in beauty.

SEMI-GLOSSary

active runway—has mowers, pickups, heifers.
adverse yaw—wife who won't fly.
aerobatics—contest between pilot and aircraft designer.
aeronautical charts—large valuable maps left at home.
Aeronca—Veronca's plain little sister.
agricultural pilot—cropduster in a clean shirt.
agonic line—wounded king of the beasts.
ailerons—what Glenn Curtiss had to invent to avoid flying a Wright.
airfoil—crew captures hijackers.
airframe—conspiracy by two or more people to lease an airplane.
Aiport Traffic Area—a triangle cornered by Washington, New York, and Chicago.
airspeed—pilot with insufficient oxygen.
airspeed indicator—pilot laughs a lot.
airstrip—nude flying.
Air Traffic Control Center—loan committee of the bank.
airworthy—how much a bank will loan you.
altimeter—big clock, always being reset by pilot.
angle of attack—angle of wing loosely connected to pilot.
angle of incidence—angle of wing firmly bolted to airplane.
Approach Control—can fly airplane without seeing it.
artificial horizon—clock that always reads about 9:20.

SEMI-GLOSSARY 195

ATC—Germanic exclamation. Originally "Ach!"
ATCC—same, said with dentures.
ATIS—Germanic, ATIS in Vunderland.
ATP—place where Apaches sleep.
Automatic Direction Finder—also called Interstate Highway.
autopilot—pilot who flies just like he drives.
avionics—a birdland.

bank—the real AOPA.
barnstormer—any pilot over 60.
base leg—stewardess with a deep voice.
Beechcraft—is to Wichita what Benz is to Stuttgart.
biplane—no sexual preference.
booster pump—verification of what maker thinks of his fuel pump.
bush pilot—student in a Cessna 152 on approach to JFK at 6:00 P.M.
buzzing—inviting your girl to your funeral before the wedding.

camber—shape of wing before riveting.
carburetor heat—small knob for reducing power on takeoff.
cardinal altitudes—infallible altitudes, decreed by the Cardinals and the Pope in Oklahoma City.
ceiling—last layer before the floor.
CFI—abbreviation of "Go on, see if I care!"
Cessna—the Chevrolet-six of aircraft.
Champ—see **rag doll**.
check ride—guilty until proven innocent.
checklist—used only by students and airline captains.
chocks—slang, passenger up-chocked.
clear—what a pilot shouts when emerging from clouds.
clearance—warning sign to pilots on Interstate overpasses.
cockpit—space left over after everything else was installed.
compass rose—legendary Japanese geisha girl.
constant-speed propeller—a shiftless Hartzell.
Continental Control Area—owned by God, claimed by the FAA.
control tower—tallest building on airport, filled with government employees.
control wheel—replaced joystick so as to have a place to mount the clock.
Control Zone—FAA-designated near-miss area.
controlled airport—there is no such thing.

convection currents—barf bag opener.
coordinated flight—when distance and fuel come out about even.
co-pilot—last form of serfdom in Western world.
course line—drawn with a thick pencil.
course deviation indicator—called "left/right needle," except in presence of FAA.
cowling—a young cow.
cross-controlled—Bonanza pilot in wrong seat.
crosswind—creates cross pilots.
Cub—more fun than a bear.

dead reckoning—when fuel and distance do not come out even.
dead stick—an old barnstormer.
density altitude—widely recognized excuse for failure to be able to fly before hitting the fence.
designee—U.S. Government employee who is not paid by the U.S. Government.
deviation—an unnatural act between consenting compasses.
dew point—precedes *don't point* by five degrees Fahrenheit.
directional gyro—an instrument that can invent compass headings.
DME—a very expensive instrument, located in the repair shop.
dive—an establishment near an airport, catering to aviators.
downwind leg—stewardess in split skirt.
drag—pilot wearing leather jacket, scarf, boots, helmet, and goggles.
drift—automatic flight extender.
dual controls—pilots equally armed.
dual instruction—one yelling, one sweating.

elevators—misnamed speed control device.

FAA—fast-growing organism with many arms and no head.
FAA examiner—good ol' boy who works for a mean company.
FARs—set of books used to ground airplanes and pilots.
fin—a place to hang the rudder.
final approach—single-engine, single-pilot, night IFR below minimums.
final fix—being out of gas, altitude, and ideas all at the same time.
FBO—to go broke over an extended period of time.
flame out—a divorce over flying.
flaps—how a helicopter flies.

flare out—lit cigar dropped in cockpit.
flight controller—like birth controller, a pill taken to prevent unwanted surprises.
flight line—act of selling an airplane.
flight plan—a small piece of paper that will support an airplane in court.
Flight Service Station—a radio talk show being replaced by automation.
flying speed—just above *falling speed*, important to know.
forced landing—there is no such thing. All airplanes will eventually come down. They do not need to be forced.
fuel selector—small hidden valve used for shortening flight.
fuselage—the last part added after airplanes were flown.

general aviation—what's left after the airlines and military.
glide—an airplane quietly seeking a state of being parked.
go-round—a surprise party.
gross weight—basis for all lying done in flying.
Ground Control—person who owns the airport.
ground loop—precautionary 360 degree scan of airport after landing.
ground school—place to learn that the ground is stronger than the airplane.
ground speed—most disappointing thing about flying.
gust locks—high-altitude pilot after Mexican dinner.
gyro—aviation word for Lorelei.

hangar—a huge bird's-nest, floored with airplanes.
heading indicator—small town with big water tank.
Hobbs meter—an instrument connected to pilot's wallet.
hood time—mechanic trying to get cowling off a Mooney.
horizontal stabilizer—a small wing whose location on the airplane is still in active dispute.
hypoxia—the state of being higher than you think you are.

ident—line boy who admits to hangar rash.
IFR—absurd flight when one can't see out the windows.
instrument pilot—can disappear at one airport, re-appear at another.
isogonic line—king of the beasts in mythological land of Isogon.

joystick—controlled aircraft before two-way radios.

Kollsman window—on-board bifocal tester.
knots—see **nautical mile**.

landing light—the state of arriving with no passengers, baggage, and little fuel.
leading edge—precedes the trailing edge in orderly flight.
lift—what a pilot needs next after he's found the men's room and the telephone.
locator outer marker—assists FAA in locating crash site.

mag check—a written promise to pay from Maggie.
magneto—a primitive ignition system discarded by the auto industry early in 20th century.
Mayday—old European rallye call to announce start of the festivities.
mixture control—small red knob for making cabin quieter.
monoplane—any unmarried airplane.
Mooney—an airplane that can be parked anywhere on a ramp full of Cessnas.
multiengine—cost of everything is doubled.
mushing—old barnstormer with no teeth.

nautical mile—for use with knots.
NDB—slang for straight flight, as in *NDB line*.
non-precision approach—pilot lands at wrong airport.
nosewheel—flimsy device to keep propeller out of pavement.

omni—small radio station with only one program.
OBS—a flying physician: Obstetrics, Beechcraft, and Surgery.

P-factor—lightplane with six-hour endurance.
pilotage—sixteen or older.
Piper—Pied, who led all the children to flying.
pitot—will catch bugs, but only one at a time.
power setting—complex of government buildings at Oklahoma City.
prime—has college degree, 1000 hours turbine time, and is under 30.
private pilot—a species like the bison, whooping crane, and so forth.
propeller—called *airscrew* in England, but not here.

rag doll—a small, old airplane, covered with faith, hope, and cotton.
relative wind—mother-in-law seated directly behind pilot.
retractable landing gear—a device like the toilet tank flush mechanism, not yet perfected.

rich—home in the country, Cub in the barn, 2,000 feet of private runway.
roll—what it takes to own aircraft radar.
rudder—device for steering boats.
runup—non-mechanic making mechanical judgments.

snap roll—fortune-cookie single-engine takeoff in twin.
solo—insufficient altitude, hits fences, shrubs, etc.
span—meat-like substance often found in on-board lunches.
stabilator—like *cafetorium* or *metroplex*, not a real word.
stall—American slang for *finis*: "Stall she wrote."
struts—what pilot does after low-minimum IFR landing.
student pilot—lovable person with no money.
supercooling—pilot does not own the engine.

TACAN—bird with a big bill, like the DME.
taildragger—passenger of over 250 pounds.
takeoff—what an airplane will do to avoid further punishment on the runway.
taxi—art of traveling in a 35-foot-wide tricycle.
TBO—short for *time before oh oh!*
TCA—private sky owned and operated by the FAA, unfortunately always located near airports.
threshold—to get a firm grip just before landing.
throttle—to grip student by throat.
torque—Writs and Torques, legal terms.
touch-and-go—airline pilot with wife at each end of his route.
traffic pattern—FAA-designated collision points.
transponder—you trans, they ponder.
trim tab—a sugar-free control surface.
T-tail—fashion designer influence on light aircraft.
turbine—airplane that flies very fast to escape fire.
turboprop—turbine aircraft slowed down with propeller.

uncontrolled airport—propaganda phrase to describe meadow with small airplanes coming and going.
UNICOM—a mythological beast that speaks through one horn in the middle of its forehead.

vectors—a board game like chess, played in the sky.
VFR—if you have it, you don't need it.

VOR—single aiming point for oncoming aircraft of all sizes and speeds.

wake turbulence—rowdy funeral for bold pilot.
walk-around—chance for pilot to go pee behind airplane.
weight & balance—seldom-used fortune-telling by numerology.
wind sock—an old barnstormer.

yaw—expression of Southern hospitality, as in "Yaw come back, heah?"
Yankee—what Yankee pilots think is phonetic for Y.

Zulu time—meaning what time it is in Greenwich or wherever you are, but not in Zulu. A simplification.